AF531384

# ADVANCED NANOTECHNOLOGY

ENCYCLOPAEDIA OF NANOSCIENCE - I

# ADVANCED NANOTECHNOLOGY

*By*

**Dr. S.K. Prasad**

*School of Studies of Zoology & Biotechnology*
*Vikram University*
*Ujjain (M.P.)*
*(India)*

**DISCOVERY PUBLISHING HOUSE PVT. LTD.**
**NEW DELHI-110 002**

First Published - 2008
Reprinted - 2017

ISBN: 978-81-8356-247-8 (Set)
ISBN: 978-81-8356-292-8



**Advanced Nanotechnology**

*Published by:*

**DISCOVERY PUBLISHING HOUSE PVT. LTD.**
4383/4B, Ansari Road Darya Ganj
New Delhi - 110 002 (India)
Phone: +91-11-23279245, 43596064-65
Fax: +91-11-23253475
*E-mail:* discoverypublishinghouse@gmail.com
sales@discoverypublishinggroup.com
*web:* www.discoverypublishinggroup.com

*Printed at:*
Infinity Imaging Systems
Delhi

# PREFACE

There has been rapid progress in nanoscience over the last few years, particularly in the area of miniaturization. The present title **Advanced Nanotechnology** is the first scientifically detailed description of developments that will revolutionize most of the industrial processes and products currently in use. This ground breaking work draws on physics and chemistry to establish basic concepts and analytical tools and thus only it provides an indispensable introduction to the emerging field of Nanoscience, Scientific research is the ultimate tool in pushing forward the limit of understanding. But, as with any tool, research is only powerful if used properly, and to its full effects. The more we learn, the more we discover connections threading through biochemical and biophysical world. In writing this Encyclopaedia the author has made every effort to present these connections in a way that will help first time students of Nanoscience understand the subject and how very relevant it is to their lives.

In the preparation of this book large number of books and research papers have been consulted. So no authenticity is claimed.

The author expresses his gratitude to Mr. Wasan and staff of M/s Discovery Publishing House for their whole hearted co-operation in the publication of this book.

The author tried hard to be accurate and upto date in statement and realises the impossibility of completely avoiding errors therefore, the author will greatly appreciate having his attention called to any questionable statement.

**Author**

# CONTENTS

# 1

# INTRODUCTION

These are extraordinary times. And we face an extraordinary challenge. Our strength as well as our convictions have imposed upon this nation the role of leader in freedom cause. No role in history could be more difficult or more important. We stand for freedom ... The great battleground for the defense and expansion of freedom today is the whole southern half of the globe—Asia Latin America, Africa and the Middle East.... Yet [our opponents] aggression is more often concealed than open. They have fired no missiles; and their troops are seldom seen. They send arms, agitators aid, technicians and propaganda to every troubled area. But where fighting is required it is usually done by others—by guerrillas striking at night, by assassins striking alone assassins who have taken the lives of four thousand civil officers in the last twelve months in Vietnam alone—by subversives and saboteurs and insurrectionists, who in some cases control whole areas inside of independent nations.

The reference to Vietnam is the only thing that differentiates this portion of John F. Kennedy's 1961 State of the Union address from one that might be heard today. Since the 1960s the nature of war and of national security has changed drastically. JFK sought a way to address a world where terrorism and unconventional warfare were the rule and not the exception. America had—and still has—the strongest conventional and nuclear military in the world, but JFK realized that its ability to address these new threats was limited. He requested funds for a variety of programs to increase national security Perhaps the most long lasting of these programs was the further establishment and expansion of the Special Forces, which have been a keystone of America's strategy in every conflict since.

Two years before Kennedy's address, in California, Richard Feynman, a Nobel Prize—winning physicist and one of the century's great scientific visionaries, was giving a very different speech: I would like to describe a field in which little has been done, but in which an enormous amount can be done in principle. This field is not quite the same as the others in that it will not tell us much of fundamental physics but it is more like solid-state physics in the sense that it might tell us much of great interest about the strange phenomena that occur in complex situations.

Furthermore, a point that is most important is that it would have an enormous number of technical applications, What want to talk about is the problem of manipulating and controlling things on a small scale. As soon as he mention this, people tell him about miniaturization, and how far it has progressed today. They tell me about electric motors that are the size of the nail on your small finger. And there is a device on the market, they tell me, by which you can write the Lord Prayer on the head of a pin. But that nothing; that the most-primitive, ha king step in the direction he intend to discuss. It is a staggeringly small world that is below. In the year 2000, when they look back at this age, they will wonder why it was not until the year 1960 that anybody began seriously to move in this direction.

Feynman went on to lay out his vision for this entirely new field. He described computers that could adapt to changing circumstances, primitive artificial intelligence, and data storage that could put the contents of every book in the California Institute of Technology library onto a single library card. Moreover, he spoke of dozens of applications for this new field that hinged on the ability to manipulate mat at its ultimate scale, atom-by-atom and molecule-by-molecule This was in the days when vacuum tubes were common, and computers were the size of a whole room yet less powerful than a modern cellular phone.

The challenge was not taken up immediately. Instead, miniaturization continued to take its "primitive halting steps" for ward toward the information technology revolution in the 1990s. It wasn't until close to the year 2000 that new tools to see and work at the level of atoms and molecules became widely available. At around the same time, it became clear that microchips could not continue to be improved by evolving current techniques. Feynman's science of the very small suddenly became both possible and necessary. It became known as *nanoscience*, and its applications became known as *nanotechnology*.

On September 11, 2001, America became the victim of the most violent terrorist attack it had yet endured—an attack carried out by fewer than 20 men using knives and box cutters killed more than 3,000 people in New York, Pennsylvania, and Washington, D.C. The attacks broke the nation's heart and forced its citizens to reconsider many of JFK's observations. How can a country protect itself against forces that, while possibly state sponsored are not the troops of a sovereign nation? How do we protect our children, airplanes roads, schools, buildings, and infrastructure against the threat of terrorism?

Terrorism was not new to America. The World Trade Center had been bombed before, Pan Am flight 103 was hijacked Al Capone bombed businesses that would not pay protection money, the Unabomber sent explosives to leading scientists, and Timothy McVeigh attacked the Oklahoma City Federal Building. Nonetheless, an attack of the size and scope of' 9/11 carried out just as America's economy was entering a recession after its euphoric "tech bubble" years—resulted in a great awakening to the fact that all was not right in the world and that we were not, even on our home soil, anywhere near so safe and omnipotent as we had thought.

The 2003 war in Iraq has brought these points into even sharper focus. If terrorists or countries sponsoring terror also possess chemical and biological weapons, what can we do to defend our troops and our cities? If our only mode of response to a terrorist attack is to invade an entire nation, how do we protect our people at home from revenge attacks, and how do we assure the safety of innocent civilians in the countries we invade? How do we work quickly and cleanly to avoid having matters spiral out of control the way they did when a terrorist's bullet started World War I? And finally, how do we find and stop the individuals, such as Osama bin Laden or Saddam Hussein, who are most responsible?

The government's initial responses to these problems—the formation of a unified Department of Homeland Security; a new Transportation Security Administration; increased funding for local police, fire departments, and other "first responders"; and a renewed emphasis on intelligence gathering and interagency coordination—were important but not sufficient. Faced with the grim possibilities of chemical or biological attacks and with the increased threats of terrorism brought about by the wars in Iraq and Afghanistan, some of these multibillion dollar agencies did little more than increase the national threat advisory level, recommend limited inoculation for some common bioagents such as anthrax and smallpox, and suggest that citizens purchase duct tape and plastic sheeting to create a protected room at home.

None of these precautions were terribly helpful or workable. Few civilians understood the obscure colour-coded system of the national threat advisory level, and they were given no practical recommendations for heightening their security when the level was increased, other than to curtail travel and other unnecessary activities, such as shopping—something which did not help a depressed economy. Sealing rooms with plastic sheeting and duct tape would do little or no good in the case of attack, since any room that will seal out toxins will also seal out breathable air. Inoculation regimens for certain biological agents made a bit more sense, but there are so many different biotoxins and biotoxin strains that inoculating against all of them is impossible.

Finally, renewed intelligence efforts have certainly helped in catching terrorists, but they also resulted in policies such as the USA Patriot Act and the proposed Total Information Awareness Act that may destroy the very freedoms and liberties we are trying to protect. As Benjamin Franklin, scientist and politician, remarked, "They that can give up essential liberty tò obtain a little temporary safety deserve neither liberty nor safety." In the battlefield, the situation is little better than in the homeland scenario we have just discussed. When facing the threat of chemical or biological attack, soldiers had either to wear thick chemical defense suits, which were similar in weight and comfort to SCUBA suits and made desert fighting next to impossible, or to take their chances in the face of these serious threats.

While the search for long-term solutions to these problems continues, what America and all the other nations of the free world need is an arsenal of "silver bullet" technologies that address these very specific needs. The military needs light weight uniforms that allow troops to fight in the desert while protecting them from bullets, shrapnel, chemical gases, and biological toxins. It also needs better smart munitions; improved stealth features; tougher, lighter vehicles and battle gear; more unmanned combat capabilities; and better battle field information systems that emphasize smaller units well inside hostile territory. UN inspectors need to be able to find and trace weapons of mass destruction quickly and efficiently. Water treatment facilities, mail sorting rooms, and transportation centers need sensors that can find explosives and other dangerous cargoes immediately, in small quantities, and with out error. First responders need substances that can neutralize chemical and biological warfare agents. Doctors need treatments to help those who have been exposed. Industry needs more secure and robust communications and computer systems. Law enforcement agencies need better capability to identify, intercept, and decode information from potentially hostile groups. And everyone needs new economic drivers and alternative sources of clean, cheap energy that will solve environmental problems and reduce our dependence on the foreign oil that so often comes from troubled parts of the world.

All of these things and many, many more are being made possible by nano-science and nanotechnology, converging the visions of Feynman and JFK. These sciences have prompted Clifford Lau, director of the Office of Basic Research at the U.S. Department of Defense, to say that nanotechnology will eventually alter warfare more than the invention of gunpowder. While policy and the people in all the uniformed services are crucial to security, without the tools made possible by nanotechnology, the world may not be much safer than it was before September 11th With the first wave of intense fighting in Iraq and Afghanistan over the urgency of these issues may seem to have diminished. It has not. Complacency in the wake of victory is among the most dangerous mistakes to make in war. Security is a long-term problem, and though it may not be the sole obsession of most citizens (as it has been during most periods of human history), it remains a great concern and the first duty of government.

## Case for Nanotechnology

Nanotechnology has immediate applications for defense and security, and they are the focus of this book, but it also has broader possibilities that will eventually touch every aspect of our lives. Many of the applications discussed here will have a profound impact on how we deal with disease outbreaks such as AIDS or SARS. Energy production, environmental repair and remediation, and advanced computing are also so-called dual-use applications, where industry and the military can pool resources for mutual benefit. There is also a host of completely nonmilitary applications—nanotechnology is already used for making better clothing, sports equipment, and treatments for diseases from diabetes to cancer—but much of that is beyond the scope of this book. Still, the applications abound. As Shimon Peres, former prime minister of Israel, observed, "That which has been achieved by the atomic bomb in the field of military strategy will be accomplished in the future by nanotechnology in the field of civil potential."

Unfortunately, technology is often a double-edged sword. The cellular phones and airplanes that are so much a part of everyday life were also necessary elements to the terrorists' plot on 9/11. Nanotechnology could be the same. Never before has a single branch of technology promised to be so transforming or powerful, nor has a technological revolution happened so quickly. It is of enormous importance that we examine both the promise and perils of nanotechnology and that we embark on this task quickly, with our full attention. With the first nanotechnology products already appearing on store shelves and on the battlefield, we cannot afford to wait.

There has been a great deal of hype about nanotechnology. Enthusiasts say that it will solve all the world's problems, from ending hunger to achieving eternal life. Critics cite concerns ranging from regulatory and intellectual property issues to the idea of "gray goo," usually meaning nanotechnology-enabled, self-aware, self replicating machines that would ultimately take over the world. Different views of nanotechnology inform movies, TV, and science fiction from Star Trek and the X-Files to Michael Crichton's thriller Prey. Almost none of these portray what nanotechnology is really about. Nanoscience is not the science of creating invisible robots to assemble kitchen sinks out of the air or to invade a human body. It is the science of utilizing the unique properties of the design scale of nature to make materials and perform functions that could not be made or performed any other way. While it might be exaggerating to say that nano-bots (those tiny, invisible robots) are absolutely impossible, there is a compelling body of evidence showing that their near-term development is extraordinarily unlikely.

Criticism of nanotechnology also takes another form. Many observers point out that while there has been a great deal of hype and a reasonable amount of investment in nano-science and nanotechnology (governments and private firms world wide spent some $4 billion on these fields in 2002, though less than half of the total was spent in the U.S.), there are few nanotechnology products to justify the excitement. This is unfair and a bit premature. While $1 billion may sound like a great deal of money, it is widely distributed for a host of new applications, and it is less than the military often spends to develop a single new weapons system. (Ballistic missile defense, for example, consumes some $8 billion per year with far fewer result.) It is less than 3% of the cost of the war in Iraq. And it is smaller than the internal R&D budgets of some individual pharmaceutical firms. In military development terms, nano technology still has a very small budget and a small program, but it is growing rapidly.

There are other flaws in the argument that nanotechnology is more hype than substance. Most of the dollars spent to date have been on fundamental research, and it may take several years to develop the results into useful products. Also, some heavy-hitting, nanotechnology-based products are already on the market: zeolite technology for oil refining currently saves the U.S. some 400 million barrels of oil a year, and GMR (*giant magneto-resistance*) technology is what makes multi-gigabyte computer hard drives possible. Since not all nano technology products have labels saying "Nano Inside," it is sometimes difficult for consumers to know which products are nano-enabled and which are not. However, with the U.S. defense budget above $400 billion per year, and the combined market impact of global defense and energy at more than $4 trillion, if nanotechnology achieves its expected impact over the next 10 years in these sectors alone, the National Science Foundation's estimate that it will be a $1 trillion industry by 2015 could, if anything, be rather low.

This book examines some of the more immediate and key areas of nano-science and nanotechnology as they apply to homeland security and to national defense. It emphasizes the capabilities of these new fields and makes some recommendations for integrating them into policy while avoiding the perils they pose. There is also a significant section on the ethical aspects of nanotechnology. The book is intended to be almost entirely nanotechnical except to the extent that is necessary to explain the basic ideas.

# 2

# NANO REVOLUTION

In a cage in front of us is a pregnant mouse. To one side, two women are arguing about the meaning of a live vide picture that shows molecules interacting in a cell inside the mouse. The pregnant mouse is cleaning her fur and occasionally scratching at a while thread that emerges from the side of her abdomen. The thread enters the skin surprisingly cleanly, because its surface exhibits mouse skin-cell markers and is accepted as normal by the surrounding tissue. The thread travels under the skin and then dives into the muscle of her abdomen, again accepted as normal tissue by the mouse's cells. Though the thread is small in diameter, it has a complex structure of fiber optic links and power cables. The thread eventually crosses into the womb of the pregnant mouse and into the circulatory system of a mouse embryo. In the thymus gland of the embryo, the thread is firmly anchored to the wall of a blood vessel and begins a succession of narrowing until it is no wider than a blood cell. The tip of the thread is just withdrawing from an *immune system* T cell and poised to enter another cell. The tip left no wound when it withdrew, for its surface carried the same phospholipid pattern as the membrane of the cell. As it enters the next cell, it becomes an integral part of the cell's membrane. As before, the thread is accepted by the membrane in the same way that the membrane accepts large proteins that also cross it in their role as molecular pumps.

How is this thread accepted inside the body of an animal without rejection? The side of the thread consists of a series of flaps that can lift to expose a variety of different manufactured surfaces. The tip of the thread is an octopus-like array of sensors on stalks. Each is no thicker than a large protein molecule and waves constantly, probing its watery surroundings. At the base of each stalk, motors whir to generate movements, and simple mechanical "*electronics*" collect the data from the sensors. The sensors "*feel*" their environment using force, electric charge, moving magnetic fields, and a variety of hydrophobic and hydrophilic probes. The sensors are small-scale version of the rich family of scanning-probe microscopes. The data are sent back up the fiber optic channels in the thread for processing by computers in the lab.

The thread and its accompanying computer and displays form in-vivo *nanoscope*, a vital tool for the hands-on-investigation of biological processes inside the cells of a living animal. In the room with the mouse cage, the scientists argue about the process by which the T-cells are being "*educated*" in the thymus of the mouse embryo. On a dozen computer screens around the room, live video pictures display false-colour views of what the tip of the thread is sensing inside the mouse cell. Just as Landsat images are often displayed, each monitor combines the data from several sensors into a single colour picture. The colours are not "*true*" because the sensors do not see light. Instead, a computer colours the pictures "*falsely*" in order to highlight interesting details.

The two women are doing research systems. It has been known for quite some time that immune system cells have to be educated about what is the body's own tissue and what is foreign matter that should be destroyed. But the exact mechanism for how this happens in the thymus gland is not well understood. If the details of the ways in which the body's T cells distinguish between friend and foe were known, the secrets of arthritis, diabetes, and other autoimmune diseases might be unlocked.

Before the development of the in-vivo nanoscope, research on the chemical processes inside living cells was extremely difficult. What scientists learned in the days before the nanoscope is amazing, considering how indirectly the knowledge was obtained. To study a process inside a live cell, one often had to fuse that type of cell with cancer cells so they would grow without restriction in the lab. Then, while keeping finicky cell cultures, alive, one had to introduce dyes or radioactive chemicals that would bind to the portion of the cell under scrutiny. Getting all these indirect methods to work required a good idea of what one was looking for. It was like drilling a hole in a bank vault and using a piece of wire to poke around in order to find out how the lock worked, or deducing the layout of the complex gears inside without being able to see them.

In the lab, with the mouse, the two scientists continue to argue, "well, if that's true, then we should seep piece of DNA actually being spliced here, but we're not." "You just aren't looking in the right place. Look over there." The first scientist looks at several false-colour displays and concentrates on moving the controls for a minute. "There–look at that." "Yes, but how do not know that that's really what we're seeing? Here, let's put some better filters on that image ..."

With the direct visualization made possible by the in-vivo nanoscope, biologists can learn in an afternoon what formerly took years of grouping with frustratingly indirect methods. A new generation of drugs and therapies is emerging as a result of this leap in understanding of the process within normal, living cells.

## Nanoscope

The *in-vivo* nanoscope is the product of an awkward stage in the development of nanotechnology. It required a building full of equipment and graduate students to construct the nanomachines at the tip of the nanoscope. Most of the "*thread*" and its surface are made with conventional technology. The computers with the intelligence for processing the images are conventional desktop machines sitting in the lab at the other end of the thread. True assemblers have not yet been made.

*Assemblers*, universal nanomachines, are nanoscale machines that can place atoms with precision and bond them in any possible pattern by following a stored design, or program. At the time of the invention of the in-vivo nanoscope, a building full of equipment can make a nanomachine, but no existing nanomachine will immediately be capable of making another nanomachine by itself. A true assembler must contain a computer, it must be able to "*see*" and make sense of its surroundings, it must be able to propel itself and navigate, and it must be able to handle and bond atoms of various kinds. Assemblers will be built someday, but not in the same decade as the simple nanomachines located at the tip of the *in vivo* nanoscope.

## "Two-Week Revolution"

Never in history has a technology been so clear in our vision yet so frustratingly out of reach as in nanotechnology, but we can not build them now. If we had a single assembler, we could order it to build another one. Those two could produce more, until any desired number were produced. Then we could tell them to build other things. A great flourishing of new nanotech machines would follow—as fast as we could design them.

But there is a technological bottleneck before we can build the first assembler. We have been accumulating plans for pieces of nanomachines, starting with the first bearing designs presented in 1987, but we remain unable to build them. It may be decades before we can actually build the first working assembler. In that time, we will accumulate many levels of designs, leading to a spectrum of useful nanomachines. When the first assembler works, a stampede of working machines could follow. All we have to do is dust off the designs and tell the newly created assemblers to go to work.

This sudden surge of working nanomachines has been called the "*two-week revolution.*" In the first two weeks after the assembler breakthrough, the world may change radically. For some this is not a metaphor but a prediction of great change in a matter of day. Entire new systems of fully functional technology will emerge, ready to transform the world.

The two-week revolution will not happen that way. Development will occur gradually after the first assembler works. Using the development of computers as a parallel, we can see that design and construction that must go hand in hand. If we accumulate a backlog of designs, they simply will not work at first. Steady and magnificent technological progress has been the hallmark of the last half century. It is a testament to the power of the product development cycle: design something, test it, learn what needs to be changed, modify the design, and repeat the cycle. The learning curve for computer chips shows that the price of a given chip falls in relation to the total number of units shipped since the product first came out. The ability of an industry to learn how to make a product depends very strongly on how many units have been made. Hundreds of little corrections and small tricks help lower the price of a unit in the next production run. Sheer experience is the most important factor. Many products from VCRs to photocopiers exhibit the same kind of learning curve and attest to the power of feedback and gradually modifying the design.

Consider the designers of a *nanomachine*. They design the device, using software on personal workstations. As soon as they think they are done with the design, they choose "*Build*" from a pop-up menu, and a group of assemblers in a tiny capsule fabricate the new device.

Suppose that the building process is absolutely perfect. The nanomachine is assembled exactly the way the blueprints say it should be. Will it work the first time?

We have a wonderful example of a perfect medium to learn from: computer software. When the software designers are finished writing a program, they ask the system to compile, and a software "*machine*" is built—and built perfectly. The only thing that can go wrong are "*bugs.*" A bug occurs when a program does exactly what the programmer wrote down, but this is not the intended result. One might think that this would not be much of a problem, but it is, software must be tested over and over in order to discover all of the subtle bugs. The length of time needed to test and debug a large software system in notoriously large.

## SOFTWARE BUGS

Why does software, the most perfectly manipulable of all media, still take such a long time to debug? It is because software is very dense with ideas. A small amount of software code can express a complex idea. The programmer weaves many such pieces into the complex fabric of a large system. There are lots of subtle interactions among the parts of the system. There are lots of subtle interactions among the parts of the system. Also, there is a great contrast between the static, written form of the program and the active movement of data in the running system. Programmers have to imagine what will happen dynamically when the computer executes the program. Often they picture it incorrectly. They do not see the consequences of the many threads interacting. They do not see the consequences of the many threads interacting the different combinations.

As with any complex system, the same bugs and misunderstandings face the designers of a *nanomachine*. Assuming the design is executed accurately, untested nanomachiens will have the same kind of bugs as untested software. (Not counting the fact that a nanodevice will contain a large amount of software running on its own on-board computer.) As with software development, the major problems will be caused by misunderstandings on the part of the designer. One misreads the specification of a module that someone else designed and tries to use it in the wrong way. One forgets an assumption made a month before—that a certain module would never be used in a certain way. One pushes a module beyond its design limits by overloading it in some way—making it drive too many other things, making it go too fast, not holding input signals steady for a long enough time, and so on.

One uses someone else's module correctly, but in a way that exposes a bug that no one has encountered before. Anyone who has debugged a computer program knows these frustrations intimately.

In addition to software bugs, there is a broad class of "*process errors*." In the first two weeks after the first assembler is built, the art of building nanostructures will not yet be flawless. The assembler may make a mistake and install the wrong atom at some location. It may not be able to reach the place where the next atom needs to go. It may be unable to make an element bond properly. The structure it is building may become unstable when only half finished, and react chemically or fold up in a strange way. All of these errors can be corrected; assemblers will eventually work wonderfully well—just not in the first two weeks. Many cycles of design and test will have to be performed by humans to get everything working.

As one thinks about this monumental "*Design-ahead*" project and tries to imagine planning for every conceivable eventuality, one comes to appreciate the power of fast turnaround. The technique of finishing a design, testing it, figuring out what went wrong, and fixing the design testing it, figuring out what went wrong, and fixing the design is a very powerful tool. One of the main reasons that Paul MacCready and his crew were able to build the first human-powered airplane is that they had quick turn around. Before the Gosamer Condor, many others had tried, but they all spent months painstakingly building a beautiful plane and then crashing it in a few seconds. MacCready designed his plane to be fixed. His plane also crashed almost every time it flew, but it could be repaired or rebuilt easily. Sometimes it flew again just twenty-four hours after a crash. This short design cycle enabled MacCready's crew to make rapid progress on improving the design.

This is also true of commercial products. Advances in camcorders, Walkmans, and personal computers have been much greater than the advances in space flight over the last ten years. This is largely because the design-cycle time for a consumer item is months, and the deign cycle for spacecrafts is closer to a decade. Yet we hear the proponents of the "*first two weeks*" saying we will be able to do it with infinitely slow turnaround, without any testing cycles before the first effective assembler is built.

## Revolutionary Argument

Those who believe in the two-weeks revolution argue that computer simulations cane detailed and accurate enough to find and correct virtually all the bugs before any working machine is built. They note that similar problems are faced by computer-chip designers. It can take up to six weeks to run a batch of wafers through a chip-making factory. Because of this time lag, it is very important to make sure a newly designed chip works before it is sent off to be built.

Chip designers use simulators to find out what the chip will do. Working with computer files that describe the layout of the chop, they submit them to programs that exact the circuit form the design. Simulators try out the circuit, and the designers compare the results to what they think the chip should do. One kind of simulator tests to see if individual transistors will work as expected. Another looks at the timing of signals traveling long distances across the chip. Another assumes that everything works on that level and tests the logic of the 1's and 0's. Another assumes that small sections of the chip work and tests if the chip as a whole produces the right answer. Such modular testing techniques might allows us to simulate all important aspects of nanomachines before they are built. Thus, when we are finally able to build them, we will have confidence that they will work as planned.

Indeed, there is a famous example of this approach working. During World War II, the Manhattan Project built a gas-diffusion separation plant in Oak Ridge, Tennessee. Its job was to separate the light isotope of uranium from the heavy one, which no plant had ever been built to do before. The principle upon which it ran was new. The gas was corrosive, and the plant had to resist it on every surface that touched the gas. In addition, the gas and the plant around it were so radioactive, repairs or rebuilding would be extremely difficult. There was not time to build a small-scale pilot plant.

Stephane Groueff describes the challenge. "*Barrier problems, pump problems, corrosion problems, cascade problems.* One of the major difficulties was the extraordinarily corrosive power of uranium gas; it ruled out the use of practically every known metal. Another one was the requirement of airtightness. Such a requirement exceeded—by far—anything that had ever been conceived before, even in small systems. To incorporate such airtightness on a gigantic scale, in piping hundreds of miles long and through two thousand cascade stages, seemed an impossible dream. The gaseous-diffusion system could be ruined by even an infinitesimal leakage. Such minute leakage, however, seemed unpreventable.... No existing valve, no welding technique and no pipe joint could guarantee such optimum airtightness." The plant had to work the first time. It did. Some chip designers have ad the wonderful experience of having a complex chip work the first time. Many others have had to test the chip, think hard about what went wrong, make changes, and submit the design for fabrication again. And again.

Projects to build large artifacts suffer from the same it-has-to-work-the-first-time problem. When you cannot build prototypes and properly test them, you have to go extraordinary lengths to make the machine work. The space shuttle flew with humans on board the first time. However, the amount of design effort, and the huge number of design change orders, attest to the difficulty of working this way. And indeed, the design of the shuttle is flawed in many ways.

## Design Rules

We know some very powerful tools for doing good design without testing it. "*Design rules*" are an especially powerful technique. A transistor in the middle of a computer chip can be varied in many ways. Different shapes of transistor, for example, will have different speed and power. But chip designers often give up that flexibility and agree to follow a set of rigid design rules. The rules only allow certain kinds of transistors, sometimes as few as three simple configurations. These transistors are slower than optimum and have a lot of restrictions on how far they can send their signals, and so on.

Why would designers give up all that flexibility and produce a design that is worse than it could be? Because they know it will work. If you were to design your own transistors, you would make mistakes. It is a complicated and specialized business. The set of official design rules are certified by the people who manage chip fabrication. When designers follow these rules, they are more or less guaranted that the transistors will work. A significant development in nanotechnology will be a good set of design rules. Such rules are rarely the best, but they can be used with assurance.

Another powerful tool is the idea of modules. A module is a "*black box*" of machinery whose inputs and outputs are carefully described. A designer can use someone else's module in his or her design without understanding how it works. The designer needs only to know how to hook it up and what it does. Cooking offers several examples. If you are preparing a recipe that calls for mayonnaise, it's very convenient to buy mayonnaise from the store, rather than making it yourself. You could, of course, prepare it from scratch, but you would need to know quite a bit about mayonnaise. It's often more practical to put that thought and effort into preparing the dish at hand. Buying the "mayonnaise module" off the shelf allows you to focus on the final culinary design.

Dividing a machine into modules greatly simplifies the design process. The "*owner*" of one module can become an expert in making it work better and an debug it thoroughly. Many people can use a well-designed module, and the number of things they need to think about is greatly reduced. The result is a more reliable overall design.

## Commercial Software

Simulation is a powerful tool, but it has its limits. Yes, it is theoretically possible to have designed an entire set of personal computer software in 1950—before the first commercial computer was shipped. It could have been all worked out on paper and in people's heads, but the number of levels of new ideas required is staggering. Today's software tools did not appear fully formed at the dawn of the computer age; their evolutionary emergence occurred over time. The advantage of having a spreadsheet on a personal computer was not obvious even after it was invented. At least one major company was offered the rights to VisiCalc, the first spreadsheet program for a personal computer, and refused. Nanotechnology's anticipated two-week revolution is akin to predesigning transistors, chips, memory systems, operating systems, computer science, several generations of programming languages, personal computers, and applications in the year 1950. We are asked to discover and appreciate the amazing utility of a spreadsheet, when no person has ever experienced an hour in front of a personal computer.

Another issue is the cost of massive design-ahead. Each layer of design that remains simulated but untested adds to the uncertainty of the device actually working. Investors will have a limit to the number of untested assumptions that can be believably piled on top of each other and may wait until the first assembler proves itself in the laboratory before they invest massive amounts of capital.

Patents are another consideration that will influence how much design-ahead is done. No one knows the extent to which designs for *nanomachines* will be patentable, and it may be hard to patent something that is impossible to build today. In general, patents on molecules are not given unless the molecule has been synthesized, however, the molecular parts in a nanomachine are much more like mechanical parts, and a very good computer simulation is arguably a working model of the actual machine. If patents are granted before it is possible to build the device, venture capitalists will be much more willing to pour money into design-ahead.

While some nanodevices will be designed before the first assembler is working, they will not be whole products ready to do things in the real world. They will not be the things described in the rest of this book. Those will come later. The sheer number of new layers of design in mature nanomachines make it very unlikely that all of it will work the fist time. The practical details of getting a design to actually work in the real world is a human endeavor. Humans need time to understand ideas, communicate, get used to new things, and get the right social organizations in place. Experience is the best teacher. While simulation can go a long way, we can be sure that the real thing will contain plenty of surprises.

## Conclusions

The two-week revolution will not happen. Two weeks after the first assembler works, it will be in the shop for repairs. And not many of the things that it built in those two weeks will work either. The pervasive use of assembles in our lives depends on the development of several new fields of study and entire new layers of infrastructure. It will be a human endeavor operating at human speeds. It won't happen without thousands of cycles of experimental feed back, and certainly not in the first two weeks.

# 3

# MOLECULAR MODELING AND REMODELING

Much of the work on molecular nanotechnology is now in the theoretical stage, which by and large means the molecular modeling stage—using computers to describe molecular machines and molecular devices. First will be the redesign of an antibody molecule, or what it is preferred to call a "*remodeling* of an antibody molecule," to bind a metal.

## PROTEIN MOLECULES

Let us start by considering the sizes of the basic building blocks of *proteins* in a series of views suggested by my colleague, Arthur Olson. In the natural environment, protein molecules are largely surrounded by water. In the water molecule there is one molecule of alanine, one of the smaller amino acid constituents of proteins. Another larger amino acid, *tryptophan*, is to the right of *alanine*. Proteins are made of a linear chain of hundreds of amino acid residues. Next is an alpha helix, one of the basic elements of secondary structure in a protein. The helix could be extended to almost any length, although it is not nearly as rigid as a diamond lattice. On the right, the large glob is a protein molecule, an anti-oxidant enzyme. It is still made up of carbon, hydrogen, nitrogen, oxygen, and a few sulfur and metal atoms.

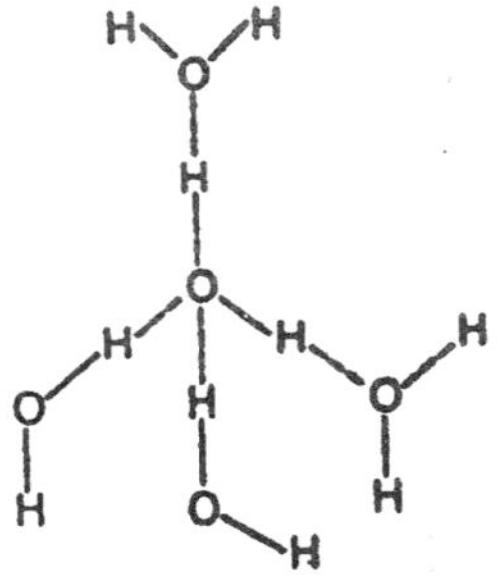

*Fig. 3.1. Water molecules.*

An antibody molecule is much larger than an average size enzyme. The antibody molecule is made of 12 domains, each of which is similar

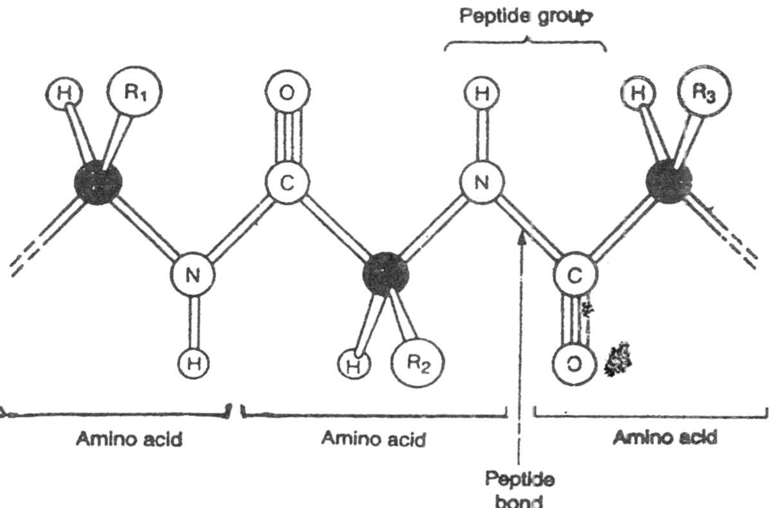


*Fig. 3.2. Protein molecules.*

in size to the enzyme. The whole antibody molecule is composed of four polypeptide chains covalently linked together, and each chain is a polymer with one of the 20 different amino acid residues at each position in the polymer chain. As far as we know, the three-dimensional shape of the protein is determined completely by the sequence of amino acid residues in the chain. However, we do not understand the relationship between linear sequence and three-dimensional structure in enough detail to be able, in general, to predict the three-dimensional structure of a given sequence. However, we do have sufficient understanding of protein structure to be able to remodel a protein of known structure to obtain a protein with designed changes in properties.

Continuing the progression to larger structures, the enzyme and antibody together with a fairly small icosahedral virus particle, having a diameter of about 60 mm The outer layer of the virus is made of protein molecules that self-assemble to form the virus structure. Under the appropriate conditions of pH, and so on, the protein molecules that form the virus shell can be mixed together, and they will self-assemble to form the virus particle.

Although the *virus* is a self-assembling molecular system, it is not a self- reproducing molecular system. It does not have the machinery to make more proteins. For that function, the virus usurps the protein synthesis machinery of the target cells to make viral proteins. Scientists

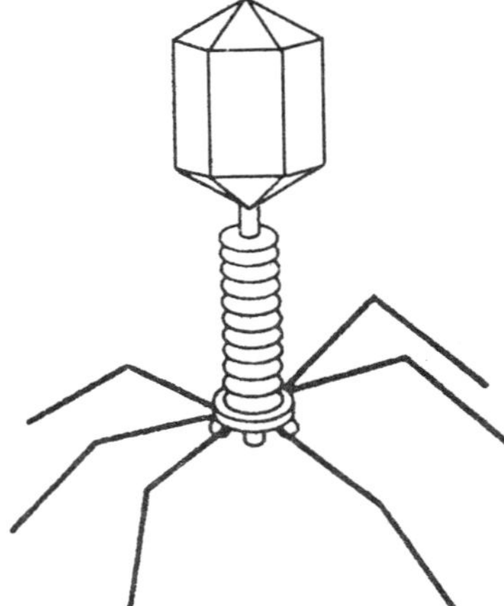

*Fig. 3.3. Bacteriophage virus.*

are also usurping the protein synthesis machinery of the cell to make proteins that have been designed or remodeled. To build new proteins atom-by-atom, biotechnologists introduce altered genetic information into cells in order to program the cells to make the new proteins.

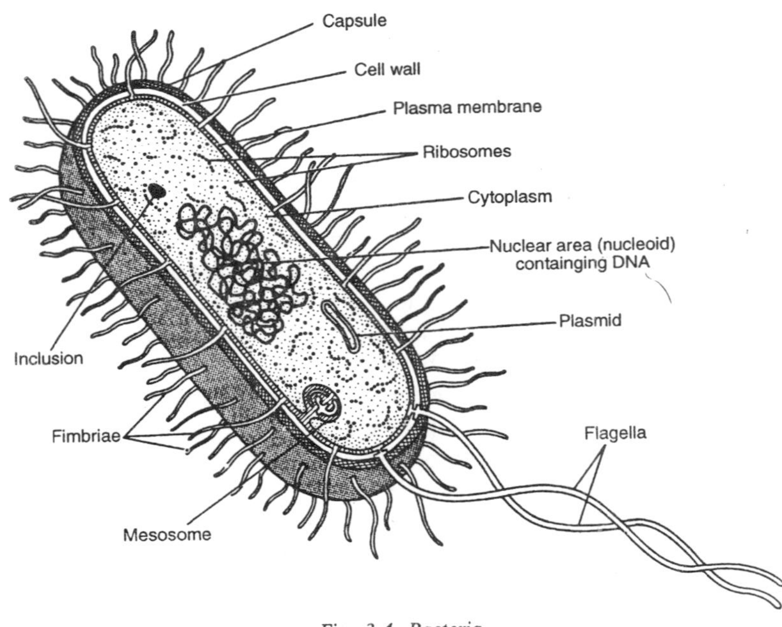


*Fig. 3.4. Bacteria.*

Still a step further up, there is a bacterial cell, the smallest truly self-reproducing molecular system. This is the celebrated *E. coli*, roughly 2,000 nm, or 2 micrometers long.

To see how atoms in protein molecules move, we have done molecular dynamics simulations on a *Cray supercomputer*. The output of such a simulation can be a video that shows, for example, a six-amino-acid residue segment with a polypeptide wiggling about, where each frame of the video represents a few femtoseconds (1 fs = $10^{-15}$ seconds).

A simulation over much longer time intervals can show the fluctuations in folding that an extended linear chain would go through while folding into the compact, three-dimensional structure of a protein molecule. Such simulations, although they are not easily verified directly by experiment, are the best models that we yet have of how a protein might look while folding. One feature that is prominent in such simulations is that a central core folds quickly, while it takes a much longer time for the outer portions of the molecule to fall into their proper places. Overall, the final structure of the protein is quite stable, although regions near the ends of the molecule remain floppy.

How can we take advantage of what we have learned from molecular dynamics simulations? The outer surface of the molecule is not shown in this representation. This representation is based upon one of the antibody structures determined by *X-ray crystallography*. The structures of about 20 different antibody molecules have been solved so that the positions of all of the atoms in the molecule are known. A comparison of these 20 structures shows that some of the positions in some parts of the molecule vary from one case to the next, but that there is a core of conserved structure that is common to all 20 structures.

## Antibody Molecule in Remodeling

This conserved core provides a framework that can be built upon for remodeling the protein. The conserved core would be maintained in the remodeling, while the parts of the antibody molecule that vary can be remodeled. The work done at Scripps changed three side chains (that is, the identities of three amino acid residues were changed), so that the remodeled protein would bind a zinc atom. It also binds copper. What this work stresses is that careful remodeling of a known structure can create an entirely new function. No natural antibody molecule has ever been discovered that binds a metal.

Many enzymatic reactions absolutely require metal ions to work. Antibodies are very specific in what they bind to. The goal of this

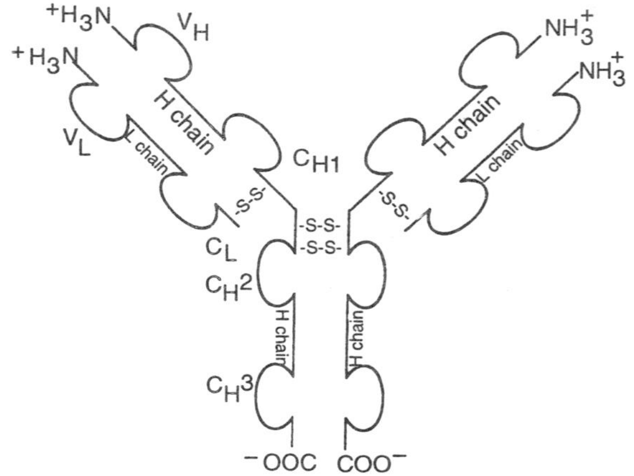

*Fig. 3.5. Structure of antibody.*

project is to be able to use the metal binding capability of the remodeled antibody molecule to catalyze reactions.

Fig. 3.5 shows two representations of the structure of an antibody molecule. One part shows the outer surface of the *antibody molecule*. The area of interest is located in the center. More frequently in our studies, we use the tubular representation of the polypeptide backbone shown in other part of the figure. However, it is necessary to be mindful of the fact that this representation shows only the skeleton of the molecule, while it is the skin of the molecule that is important in reactions because that is what the outside world sees.

To begin the remodeling work, Victoria Roberts in our laboratory analyzed the known metal-binding sites in various proteins. Different proteins bind a variety of metals, including magnesium, iron, copper, zinc, and others. The structures are from the metal-binding sites of four proteins that happen to bind zinc. Among these, the protein carbonic an hydrase was of particular interest because a computer search of protein structures showed that the structure of the carbonic an hydrase metal-binding site was very similar to a structure common to all known antibody molecules.

The basic frameworks of these two structures are very similar, but there are important differences in some of the side chains of specific amino acid residues. This structural similarity of the backbones suggested that the antibody molecule could be redesigned, replacing only those specific side chains. The hope was that the remodeled antibody molecule would retain the overall antibody structure, but that

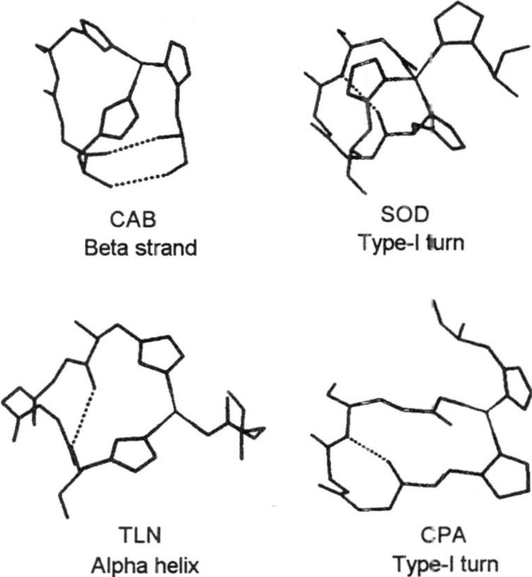


*Fig. 3.6. The metal-binding sites of four proteins. The dotted lines are hydrogen bonds.*

the altered residues would confer upon the remodeled antibody the ability to bind a metal atom. Specifically, three amino acid residues of the antibody's "*light chain*" were changed to histidines by site-

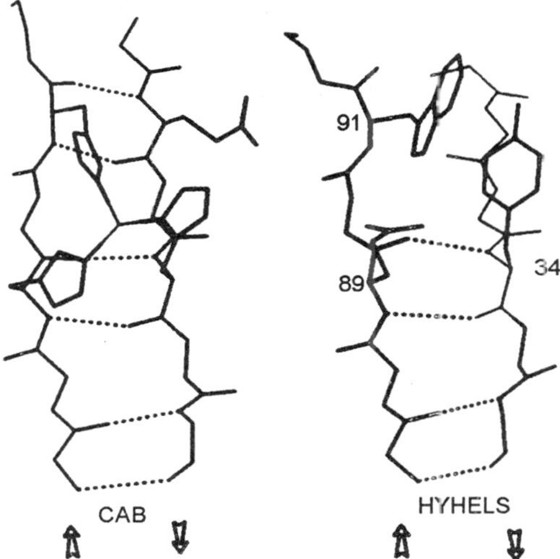


*Fig. 3.7. Structural comparison between the metal-binding site of carbonic anhydrase (left) and a section of an antibody with a similar backbone structure (right).*

directed mutagenesis, and the antibodies were expressed in bacterial cultures. When the antibodies were extracted from the bacteria and put in a solution containing copper, they did indeed bind the copper ions, with a dissociation constant of about $5 \times 10^{-7}$ M. This is, we say, within a factor of 100,000 as good as a natural metal-binding enzyme—not bad for the first attempt.

The long-term goal of remodeling the antibody molecule is that the remodeled antibody would still bind to the same target protein to which the original antibody bound, but the added ability to bind a metal atom would confer upon the remodeled antibody the ability to cut one specific covalent bond in the target protein to which the antibody bound. Antibody molecules that could cleave bonds in their target proteins would have important medical applications.

This capability to cleave a bond in the bound protein has not yet been achieved, but a project has been proposed that would involve the efforts of 10 scientists over 5 years to build such a site-specific zinc protease to cleave a specific peptide bond. The research community is making great progress toward such goals. We think that if our team does not do it, some other team will over the next few years. Nevertheless, from the standpoint of molecular nanotechnology, we are only at the point of being a few years and a great deal of effort away from the ability to cleave one specific bond in a large molecule.

## Catalytic Remodeling

An *enzyme* is a *biological catalyst*. It makes a reaction happen faster than it would other wise happen—usually millions of times faster. The molecule that we worked with was *superoxide dismutase*, an important defense that our cells have against aging and other oxidative damage. The enzyme catalyzes the conversion of two highly reactive and destructive $O_2^-$ superoxide anion radicals into hydrogen peroxide and molecular oxygen.

A computer analysis of the exact location of charged amino acid side chains in this molecule revealed that the distribution of electrostatic charge is irregular over most of the surface. However, at the active site channel, a large region of positive charge covers one-tenth of the surface area of the enzyme and serves to attract the negatively charged superoxide ion substrate.

A computer analysis of the electrostatic field set up by these positive charges shows how a negatively charged superoxide molecule is directed along a path toward the copper at the bottom of the active site channel. This channel is about 12 Å (1.2 nm) deep. The superoxide fits neatly into a pocket in the protein immediately above the copper. There are only two metal atoms (a copper ion and a zinc ion) out of more than 3,000 carbon, hydrogen, and other atoms in the protein, but they are essential to the activity of the enzyme.

The general features of the interaction of the enzyme and superoxide were apparent from early studies done about eight years ago. More recent and more accurate computer simulations of the molecular dynamics of this system use the fact that the superoxide substrates move around in the vicinity of the enzyme under electrostatically influenced Brownian motion. This newer model allows us to get numerical results that can be compared with experiment, and which agree fairly well with experiment. Over the course of the simulation, in one out o every 40 or 50 runs, the superoxide anion released in the vicinity (4.12 nm away from the center) will come into the active site channel and react with the copper In the other runs, the superoxide exits the system, which is contained in a sphere with a radius of 50 nm. Summing up over 12,000 simulated trajectories in on data point, one can calculate the reaction rate of the enzyme. These rates the can be compared to experimental data.

Our computer simulation only takes the substrate to the point of binding in the active site channel. Simulating the actual chemical reaction that happens in this site would require quantum mechanics rather than the classical mechanics upon which molecular dynamic simulations are based. However, the diffusion up to the active site is the process that largely determines the reaction rate. We can change the charge or shape of specific side chains and run the program again to see whether more or fewer of the substrates react within a given time interval.

We can look at this enzyme in more detail and see the effects of remodeling upon enzyme activity. Copper, zinc superoxide dismutase, so-called because of the two metal atoms in the active site. The structural framework of the original enzyme was maintained in the remodeled enzyme, but key side chains were altered to give a two- to three-fold increase in the rate at which the enzyme works.

The basic idea for the changes to remodel the enzyme actually came from several years of work looking at and pondering the structure of the active site channel, even before the numerical results of the more recent, more refined simulation were available. The investigators concluded that a specific network of amino acid side chains along the outside of the protein was important in stabilizing the active site channel and guiding the substrate toward the copper. That conclusion turned out to be correct because they were able to achieve the faster rates by changing these side chains in the way indicated by their interpretation.

Within the past two to three years, computer modeling of enzymes has become accurate enough to be of real help in choosing changes to make when remodeling proteins. A great need still exists for the intuitions of the investigators.

The more sophisticated computer models include the effects of the solvent (the water molecules surrounding the enzyme) upon the molecular dynamics of the protein. Water is not at all inert. It is chemically very active, as can be seen from the way in which it catalyzes the rusting of metals. Water is a polar molecule, with positive and negative charges and a complicated network of hydrogen bonds among water molecules.

The computer simulation of the enzyme would be much simpler if the enzyme could be simulated in a vacuum, but that would not give an accurate result. In order to find out how good the protein is in electrostatically guiding the substrate molecules toward the active site channel, it is necessary to extend the simulation some distance away from the protein into the surrounding solvent. The simulations had to be carried about 50 nm into the surrounding solvent, many times the diameter of the protein molecule. The result of such a simulation, which identifies the regions around the protein that are repulsive to the substrate and the regions that are attractive to the substrate.

This analysis required looking at about 10,000 molecular trajectories. It took three months to do the analysis. It was necessary to investigate the effects of different ionic strengths of the solvent, and so on. The method leaves much to be desired, even though it is the best available. It is a statistical "*Monte Carlo*" method developed by Andy McCammon and team that takes 5 or 10 minutes on the computer to follow the path of any one superoxide anion radical, and looking at only one superoxide does not tell you very much. This method is not recommended, but it can be done.

The results of computer simulation of enzyme activity and the experimentally determined enzyme activities are compared in Fig 3.8. The lower curve with the solid circles is the experimental data for the wild-type enzyme. The upper curves with the solid triangle and diamond symbols show that the remodeled enzymes are two- to threefold faster than the wild-type enzyme. The curves with the open symbols

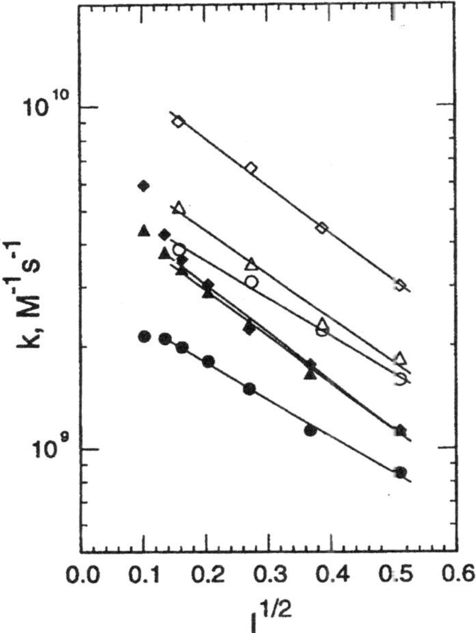


*Fig. 3.8. Comparison of experimentally determined activities (solid symbols) of wild-type and remodeled superoxide dismutase with computer simulations (open symbols) of the activities of these proteins.*

represent the best results of the computer simulation for the wild-type (lower curve, open circles) and the *remodeled enzymes* (upper curves, open triangles and diamonds).

The results show that the absolute reaction rates are not very well predicted. That is, the computer simulation predicts that both the wild-type and the remodeled proteins are substantially more active than they, in fact, are. However, the difference in reactivity between the wild-type and remodeled enzymes is well-predicted. One way of looking at these results is that these computer models are not yet good enough for doing *de novo* design of enzymes, but they do seem good enough, at least in some cases, for remodeling a protein to change its activity.

However, this remodeling process is not yet automatic. It is still true that the major benefit of computer simulation is to help enhance the intuition of the investigators, so that they can make changes in the protein and then check the effects of these changes experimentally. The simulations are on firm ground with simple compounds of carbon, oxygen, and so on. However, when you begin to deal with metals, with strained bonds, and with unusual conformations that have not been well-investigated experimentally (such as by infrared spectroscopy), then you start to get into trouble.

## STEREO LITHOGRAPHY

So far, we have been looking primarily at models that portray only the skeleton (the peptide backbone) of the proteins. We have recently been collaborating with the University of North Carolina at Chapel Hill, the University of Utah, and Allied Signal Corporation in Kansas City on a new technology (called *stereolithography*) to make physical models of the exterior surfaces of proteins. An example of such a model is an irregularly shaped piece of plastic about the size of a grape fruit that models the copper, zinc *superoxide dismutase molecule*. This solid model represents the solvent-accessible surface, as defined by Fred Richards. This is the portion of the surface of the protein that could be contacted by rolling a sphere with the diameter of a water molecule around the surface of the protein. The solvent-accessible surface thus represents the portion of the molecule that the outside world can see with which it can react.

Such a solid model is more useful to aid in visualizing what the possibilities are for reaction with this molecule. Where are the open spaces where a potential reactant can contact the protein? Looking at a framework model, by contrast, would give the false impression that the molecule is an open network containing many holes through which a potential reactant could pass to come into contact with the hollow interior of the molecule. The solid model shows that, in fact, the molecule contains very little empty space. A major advantage in helping the investigator to develop intuition is that the solid model allows the investigator to not only see, but also to feel, how the out side world experiences the molecule. In fact, the superoxide dismutase molecule forms a dimer. So, by using two solid models of the single chain subunit, the investigator can directly and manually explore the different ways in which the two subunits might interact with each other by rolling the two models over each other.

This is a particular advantage with molecular design, because molecules are irregular, three-dimensional objects (unlike most macroscopically manufactured devices) which are based upon right-angled planes. Unlike conventional machining techniques, the stereolithography equipment is not limited to any particular type of shapes. It takes the computer-produced specifications for any arbitrary three-dimensional shape and builds the specified part (whether it is a model of a protein molecule or anything else) over about an eight-hour period.

This *stereolithography technology*, developed by Three-D Systems, Inc., of Valencia, CA, has even been featured in a short newsreel video shown to airline passengers. The video showed how various small industrial products were produced as solid three-dimensional objects, guided by the output from computer-aided design programs. The newsreel video makes the point that stereolithography is actually "*three-dimensional printing.*" Actual parts are made without tooling, machining, or cutting of any kind. The parts are made from a vat of liquid plastic (called a *photopolymer*) that hardens in the presence of ultraviolet light. A laser, guided by the computer program, draws the part in the liquid plastic as an elevator inside the vat repositions the object being constructed. The object is thus assembled section-by-section, about 100 sections per inch. The advantage for an industrial design engineer is that he can now hold a physical part in his hands only hours after designing it on the a computer, rather than waiting weeks or months for it to be machined by conventional methods.

# 4

# MOVEMENT AND SELECTION OF MOLECULES

## HUMAN BODY CHEMICAL COMPOSITION

The human body consists of $\sim 7 \times 10^{27}$ atoms arranged in a highly aperiodic physical structure. Although 41 chemical elements are commonly found in the body's construction, CHON comprises 99% of its atoms. Fully 87% of human body atoms are either hydrogen or oxygen.

Somatic atoms are generally present in combined form as molecules or ions, not individual atoms. The molecules of greatest nanomedical interest are incorporated into cells or circulate freely in blood plasma or the interstitial fluid. The gross molecular contents of the typical human cell, which is 99.5% water and salts, by molecule count, and contains $\sim$5000 different types of molecules. There are 261 of the most common molecular and cellular constituents of human blood, and their normal concentrations in whole blood and plasma. This listing is far from complete. The human body is comprised of $\sim 10^5$ different molecular species, mostly proteins—a large but nonetheless finite molecular parts list. By 1998, at least $\sim 10^4$ of these proteins had been sequenced, $\sim 10^3$ had been spatially mapped, and $\sim$7,000 structures (including proteins, peptides, viruses, protein/nucleic acid complexes, nucleic acids, and carbohydrates) had been registered in the Protein Data Bank which at that time was maintained at Brookhaven National Laboratory. Given the current accelerating pace of improving technology, it is likely that the sequences and 3-D or tertiary structures of all human proteins will have been determined by the second decade of the 21st century.

Transporting and sorting such a broad range of essential molecular species will be an important basic capability of many nanomedical systems. The three principal methods for distinguishing and conveying molecules that are most useful in nanomedicine are diffusion transport, membrane filtration, and receptor-based transpor. The Chapter ends with a brief discussion of binding site engineering.

## DIFFUSION TRANSPORT

Fluidic transfer of material, known as convective-diffusive transport, can occur either by convection due to bulk flow or by diffusion due to Brownian motion. In convective transport, material is carried along fluid streamlines at the mean velocity of the fluid, with a velocity distribution such as that in Poiseuille flow. Bulk flow is customarily regarded as the most important physiological transport mechanism in the human circulation. Only for the smallest molecules, such as water or glucose, does the time required to diffuse across the width of a capillary roughly equal the time taken by a fluid element to flow the same distance ($\sim$0.02 sec). Larger molecules such as fibrinogen take $\sim$100 times longer ($\sim$2 sec) to diffuse across one capillary width.

However, bulk flow in the body is usually laminar. Transported materials travel parallel to (and thus cannot reach) fluid/solid interfaces such as the surfaces of blood vessels or membranes. Wall interactions are made possible by diffusion, a random process in which particles can move transversely to fluid streamlines in response to molecular-scale collisions.

Additionally, the movement of micronscale devices within a bulk fluid flow is dominated by viscous, not inertial, forces. Molecular transport to and from such nanodevices is governed by diffusion, not by bulk flow.

### Brownian Motion

A particle suspended in a fluid is subjected to continuous collisions, from all directions, with the surrounding molecules. If the velocities of all molecules were the same all the time, the particle would experience no net movement. However, molecules do not have a single velocity at a given temperature, but rather have a distribution of velocities of varying degrees of probability. Thus from time to time, a suspended particle receives a finite momentum of unpredictable direction and magnitude. The velocity vector of the particle changes continuously,

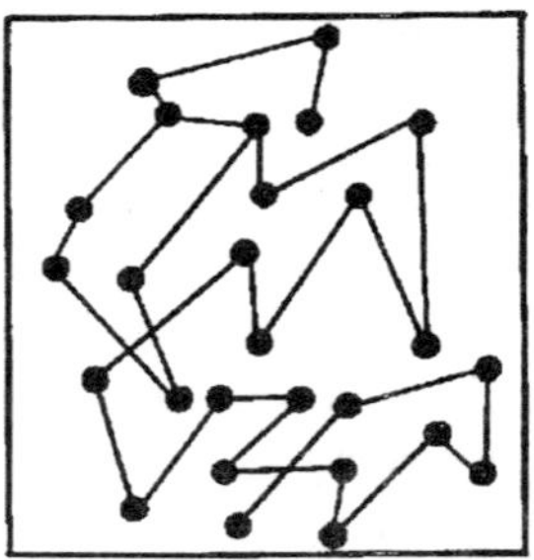

*Fig. 4.1. Brownian movement.*

resulting in an observable random zigzag movement called *Brownian motion*.

Einstein approximated the RMS (root mean square) displacement of a particle of radius R suspended in a fluid of absolute viscosity $\eta$ and temperature T, after an observation period $\tau$, as:

$$\Delta x = (kT\ \tau\ /\ 3\ \pi\eta\ R)^{1/2}\ \text{(meters)} \qquad ...(1)$$

where k = $1.381 \times 10^{-23}$ joule/kelvin (K) or 0.01381 zJ/K (Boltzmann's constant). Particles under bombardment also experience a rotational Brownian motion around randomly oriented axes, with the RMS angle of rotation:

$$\Delta\alpha = (kT\ \tau\ /\ 4\ \pi\eta\ R^3)^{1/2}\ \text{(radians)} \qquad ...(2)$$

although for $\tau < \tau_{min} = M\ /\ 15\ \pi\eta\ R$, where M is particle mass, rotation is ballistic.

In human blood plasma, with $\eta = 1.1$ centipoise ($1.1 \times 10^{-3}$ kg/m-sec) and T = 310 K, a spherical 1-micron diameter nanodevice (R = 0.5 micron) translates ~1 micron in 1 sec ($v_{brownian} \sim 10^{-6}$ m/sec) or ~8 microns (~the width of a capillary) after 77 sec ($v_{brownian} \sim 10^{7}$ m/sec), and rotates once in ~16 sec ($\tau_{min} = 2 \times 10^{-8}$ sec). In the same environment, a rigid 10-nm particle (roughly the diameter of a globular protein) would translate ~8 microns in one second ($v_{brownian} \sim 10^{-5}$ m/sec) while rotating ~250 times, due to Brownian motion ($\tau_{min} = 2 \times 10^{-12}$ sec).

The instantaneous thermal velocity over one mean free path, the average distance between collisions, is much higher than the net Brownian translational velocity would suggest. For a particle of mass $M = 4/3\ \pi\rho\ R^3$ with mean (working) density r, the mean thermal velocity is:

$$v_{thermal} = (3\ kT\ /\ M)^{1/2} \qquad ...(3)$$

At T = 310 K, a spherical 1-micron diameter nanodevice of normal density (e.g., taking $\rho \sim \rho_{H_2O} = 993.4$ kg/m$^3$ to minimize ballasting requirements) has $v_{thermal} \sim 5 \times 10^{-3}$ m/sec; for a spherical 10-nm diameter protein with $\rho \sim 1500$ kg/m$^3$, $v_{thermal} \sim 4$ m/sec.

**Passive Diffusive Intake**

Medical nanodevices will frequently be called upon to absorb some particular material from the external aqueous operating environment. Molecular diffusion presents a fundamental limit to the speed at which this absorption can occur. (Once a block of solution has passed into the interior of a nanodevice, it may be divergently subdivided and transported at ~0.01-1 m/sec along internal pathways of characteristic dimension ~1 micron far faster than the <1 mm/sec bulk diffusion velocity across 1 micron distances.)

For a spherical nanodevice of radius R, the maximum diffusive intake current is:

$$J = 4\pi \, RDC \qquad ...(4)$$

where J is the number of molecules/sec presented to the entire surface of the device, assumed to be 100% absorbed, D (m$^2$/sec) is the translational Brownian diffusion coefficient for the molecule to be absorbed, and C (molecules/m$^3$) is the steady-state concentration of the molecule far from the device.[337] (Blood concentrations in gm/cm$^3$ are converted to molecules/m$^3$ by multiplying by ($10^6 \times N_A$/MW), where $N_A = 6.023 \times 10^{23}$ molecules/mole (Avogadro's number) and MW = molecular weight in gm/mole or daltons.) For rigid spherical particles of radius R, where $R >> R_{H_2O}$, the EinsteinStokes equation gives:

$$D = kT/(6 \, \pi\eta \, R) \qquad ...(5)$$

though this is only an approximation because D varies slightly with concentration, with departure from molecular sphericalness, and other factors.

Measured diffusion coefficients in water for various molecules of physiological interest, converted to 310 K. Diffusion coefficient data for ionic salts such as NaCl and KCl, which dissociate in water and diffuse as independent ions, are for solvated electrolytes. A 1-micron (diameter) spherical nanodevice suspended in arterial blood plasma at 310 K, with $C = 7.3 \times 10^{22}$ molecules/m$^3$ of oxygen and $D = 2.0 \times 10^{-9}$ m$^2$/sec, encounters a flow rate of $J = 9.2 \times 10^8$ molecules/sec of $O_2$ impinging upon its surface. (The same calculation applied to serum glucose yields $J = 1.3 \times 10^{10}$ molecules/sec.) The characteristic time for change mediated by diffusion in a region of size L scales as ~$L^2$/

D. Across the diameter of an L = 1 micron nanodevice, small molecules such as glucose diffuse in ~0.001 sec, small proteins like hemoglobin in ~0.01 sec, and virus particles diffuse in ~0.1 sec. Diffusion coefficients of the same molecules in air at room temperature are a factor of ~60 higher, because $\eta_{air}$ ~ 183 micropoise at 20°C.

In blood, the diffusivity of larger particles is significantly elevated because local fluid motions generated by individual red cell rotation lead to greater random excursions of the particles. The effective diffusivity $D_e = D + D_r$, where the rotation-induced increase in diffusivity $D_r \sim 0.25\, R_{rbc}^{2}\, `\gamma$, with red cell radius $R_{rbc}$ ~ 2.8 microns (taken for convenience as a spherical volume equivalent) and a typical blood shear rate $`\gamma \sim 500\ sec^{-1}$, giving $D_r \sim 10^{-9}\ m^2/sec$ in normal whole blood. The elevation of diffusivity caused by red cell stirring is just 50% for $O_2$ molecules. However, for large proteins and viruses the effective diffusivity increases 10-100 times, and the effective diffusivity of particles the size of platelets is a factor of 10,000 higher than for Brownian molecular diffusion.

The diffusion current to the surface of a nanodevice can also be estimated for various nonspherical configurations. For instance, the diffusion current to both sides of an isolated thin disk of radius R is given by J = 8 R D C. The two-sided current to a square thin plate of area $L^2$ is $J = (8/\pi^{1/2})\ L\ D\ C$. The steady-state diffusion current to an isolated cylinder of length $L_c$ and radius R is approximated by $J = 2\pi L_c DC / (\ln(2L_c/R) - 1)$, for $L_c >> R$. The diffusion current through a circular hole of radius R in an impermeable wall separating regions of concentration $c_1$ and $c_2$ is $J = 4\ R\ D\ (c_1 - c_2)$.

**Active Diffusive Intake**

Foraging nanodevices operating in aqueous environments may only modestly exceed the maximum rates of passive diffusive intake by engaging in active physical movements designed to increase access to the desired molecules, as described quantitatively below.

***Diffusive stirring***

The first strategy for active diffusive intake is local stirring. For this, the nanodevice is equipped with suitable active appendages used to manipulate the fluid in its vicinity. Transport by stirring is characterized by a velocity $v_a$, the speed of the appendage, and by a length $L_a$, its distance of travel, which together define a characteristic stirring frequency $n_{stir} \sim v_a/L_a\ sec^{-1}$. Movement of molecules over a distance $L_a$ by diffusion alone is scaled by a characteristic time $\sim L_a^2/D$, which defines a characteristic diffusion frequency $n_{diff} \sim D / L_a^2$

$sec^{-1}$. Stirring will be more effective than diffusion only if $n_{stir} > n_{diff}$, that is, if $v_a > D / L_a$. For local stirring, $L_a$ cannot be much larger than the size of the nanodevice itself. Assuming $L_a = 1$ micron and $D = 10^{-9}$ $m^2$/sec for small molecules, then $v_a > 1000$ microns/sec, a faster motion than is exhibited by bacterial cells but quite modest for nanomechanical devices. With $D = 10^{-11}$ $m^2$/sec for large proteins and virus particles, $v_a > 10$ microns/sec, well within the normal microbiological range.

The ratio of stirring time to diffusion time, or Sherwood number, is:

$$N_{Sh} = L_a v_2 / D \qquad ...(6)$$

provides a dimensionless measure of the effectiveness of stirring vs. diffusion. For bacteria absorbing small molecules, $N_{Sh} \sim 10^{-2}$. Micron-scale nanodevices with 1-micron appendages capable of 0.01-1 m/sec movement can achieve $N_{Sh} \sim 10\text{-}1000$ for small to large molecules, hence could be considerably more effective stirrers.

In a classic paper, Berg and Purcell analyzed the viscous frictional energy cost of moving the stirring appendages so that the fluid surrounding a spherical object (e.g., a nanodevice) of radius R, out to some maximum stirring radius $R_s$, is maintained approximately uniform in concentration. The objective is to transfer fluid from a distant region of relatively high concentration to a place much closer to the nanodevice, thereby increasing the concentration gradient near the absorbing surface. To double the passive diffusion current by stirring, the minimum required power density is:

$$P_d = \frac{12\,\eta\,D^2}{R^4}\left(\frac{R_s + 2R}{R_s - 2R}\right) (\text{watt} / m^3) \qquad ...(7)$$

If $\eta = 1.1 \times 10^{-3}$ kg/m-sec, R = 0.5 micron, $D = 10^{-9}$ $m^2$/sec for small molecules, and using a modest $L_a = 1$ micron stirring apparatus giving $R_s = 3R$, then $P_d \sim 3 \times 10^7$ watts/$m^3$. This greatly exceeds the $10^2$-$10^6$ watts/$m^3$ power density commonly available to biological cells but lies well within the normal range for nanomechanical systems which typically operate at up to $\sim 10^9$ watts/$m^3$. For $D \sim 10^{-11}$ $m^2$/sec for large molecules, $P_d \sim 3 \times 10^3$ watts/$m^3$, which is reasonable even by biological standards. The maximum possible gain from stirring is $\sim R_s/R$, because the current is ultimately limited to what can diffuse into the stirred region.

Local heating due to stirring is minor. Given device volume V $\sim 1$ $micron^3$, $P_d = 3 \times 10^7$ watts/$m^3$, mixing distance $L_{mix} \sim 5$

microns, and thermal conductivity $K_t = 0.623$ watts/m-K for water, then $\Delta T \sim (P_d V / L_{mix} K_t) = 10$ microkelvins; taking heat capacity $C_V = 4.19 \times 10^6$ J/m$^3$-K for water, thermal equilibration time $t_{EQ} \sim L_{mix}^2 C_V / K_t = 0.2$ millisec.

***Diffusive swimming***

The second strategy for active diffusive intake is by swimming. Again, the nanodevice is equipped with suitable active propulsion equipment which enables it to move so as to continuously encounter the highest possible concentration gradient near its surface. Consider a spherical motile nanorobot of radius R propelled at constant velocity $v_{swim}$ through a fluid containing a desired molecule for which the surface of the device is essentially a perfect sink. Applying the Stokes velocity field flow around the sphere to the standard diffusion equation, a numerical solution by Berg and Purcell found that the fractional increase in the diffusion current due to swimming is proportional to $v_{swim}^2$ for $v_{swim} << D/R$, and to $v_{swim}^{1/3}$ for $v_{swim} >> D/R$.

Diffusive intake is doubled at a swimming speed $v_{swim} = 2.5$ D/R, which for 1-micron devices is ~5000 microns/sec when absorbing small molecules, ~50 microns/sec for large molecules. The viscous frictional energy cost to drive the nanodevice through the fluid, derived from Stokes' law, requires an onboard power density of:

$$P_d = 9\,\eta\, v_{swim}^2 / 2\,R^2 \qquad \text{...(8)}$$

If $\eta = 1.1 \times 10^{-3}$ kg/m-sec, $v_{swim} = 2.5$ D/R, R = 0.5 micron, D = $10^{-9}$ m$^2$/sec for small molecules, then $P_d \sim 5 \times 10^5$ watts/m$^3$. For large molecules with D = $10^{-11}$ m$^2$/sec, $P_d \sim 50$ watts/m$^3$. Thus the energy cost of diffusive swimming appears modest for nanomechanical systems; gains in diffusion by swimming for nanodevices will be restricted primarily by the maximum safe velocity that may be employed *in vivo*.

In general, outswimming diffusion requires movement over a characteristic distance $L_s \sim D/v_{swim}$. For bacteria moving at ~30 micron/sec and absorbing small molecules, then $L_s \sim 30$ microns, roughly the sprint distance exhibited by flagellar microbes such as *E. coli*. For micron-scale nanodevices moving at ~1 cm/sec, $L_s \sim$ 1-100 nm for large to small molecules.

**Diffusion Cascade Sortation**

Nanodevices may also use diffusion to sort molecules. One of the remarkable features of diffusive sortation is that an input sample consisting of a complex mixture of many different molecular species can sometimes be completely resolved into pure fractions without having any direct knowledge of the precise shapes or electrochemical characteristics of the molecules being sorted. This can be a tremendous advantage for nanodevices operating in environments containing a large number of unknown substances. Another major advantage is the ability to readily distinguish isomeric (though not chiral) molecules. As one example of many possible, molecules suspended in water will diffuse into an adjacent region of pure water at different speeds, giving rise to dissimilar time-dependent concentration gradients which may be exploited for sortation by interrupting the process before complete diffusive equilibrium is reached.

For simplicity, assume we wish to separate two molecular species initially present in solution in equal concentrations ($c_1 = c_2$), but having unequal diffusion coefficients ($D_1 < D_2$). Consider a separation apparatus with two chambers. Chamber A contains input sample concentrate. Chamber B contains pure water. A dilating gate separates the two chambers. The gate is opened for a time $\Delta t$ approximated by:

$$\Delta t = \frac{(\Delta X)^2}{2\,D_2} - \frac{L^2}{2\,D_2} \qquad \ldots(9)$$

which relates the diffusion coefficient to the mean displacement $\Delta X$, taken here as L, the length of Chamber B.

After $\Delta t$ has elapsed, the gate is closed. (A gate with 10-nm sliding segments moving at 10 cm/sec closes in 0.1 microsec.) The faster-diffusing component $D_2$ approaches diffusive equilibrium in Chamber B, but the slower-diffusing component does not; it is present only in smaller amounts. This gives a separation factor $c_2/c_1 \sim D_2/D_1$ for each diffusion sortation unit. If n units are connected in series, with each unit receiving as input the output of the previous unit, the net concentration achieved by the entire cascade is $\sim(D_2/D_1)^n$. Such cascades are commonplace in gaseous diffusion isotope separation and other applications.

Each unit consists of 5 chambers of equal volume, 7 dilating gates, 3 flap valves, 3 pistons, and 2 sieves which pass only water (or smaller) molecules. Each chamber is roughly cubical with $L \sim 35$ nm along the inside edge; including full piston throws and drives, controls, interunit piping and other support structures, each unit measures $\sim 125$ nm $\times$ 100 nm $\times$ 80 nm or $\sim 0.001$ micron$^3$ with a mass of $\sim 10^{-18}$ kg.

The following is a precise description of one complete cycle of operation for each unit:

1. The cycle begins with fluid to be sorted in Chamber A, Chambers B and W full of pure water with piston W all the way out, Chambers R and D empty with pistons R and D all the way in, and all valves and gates closed.
2. Gate AB is opened for a time $\Delta t$, then closed. For a small molecule such as urea (MW = 60 daltons), $\Delta t$ = 1 microsec; for a large molecule such as the enzyme urease (MW = 482,700 daltons), $\Delta t$ = 35 microsec.
3. Valves AI- and AI+, and gate AR, are opened. Piston R is drawn fully out, slowly to preserve laminar flow and to prevent mixing. Fluid in Chamber A is drawn into Chamber R. Fluid passing through the DO gate of the previous unit in the cascade enters Chamber A through valve AI-. Fluid passing through the RO valve of the subsequent unit in the cascade enters Chamber A through valve AI+. All valves and gates are closed. Chamber A is now ready for the next cycle.
4. Gates WB and BD are opened. Piston W is slowly pushed all the way in while piston D is slowly pulled all the way out. Concentrated solution in Chamber B is transferred into Chamber D as pure water in Chamber W is transferred into Chamber B, again preserving laminar flow. Both gates are closed; Chamber B is now ready for the next cycle.
5. Gates RW and DW are opened. Pistons R and D are slowly and simultaneously pushed halfway in while piston W is pulled all the way out. Forced at high pressure (~160 atm) through ~0.3 nm diameter sieve pores, half of the solvent water present in Chambers R and D is pushed into Chamber W, filling Chamber W with water. (This design allows for easy backflushing if sieve pores become clogged.) Both gates are closed; Chamber W is now ready for the next cycle.
6. Valve RO and gate DO are opened. Pistons R and D are slowly and simultaneously pushed the rest of the way in. Concentrated return fluid passes through valve RO and back to the AI+ input port of the previous unit in the cascade for further extraction. Concentrated diffusant fluid passes through gate DO and on to the AI- input port of the subsequent unit in the cascade for further purification. The valve and gate are closed; Chambers R and D are now empty and ready for the next cycle.

7. Return to Step (1). Adjacent units operate in counterphase while previous and subsequent units operate in synchrony, in a two-phase system.

Increasingly purified sample passes through a multi-unit sortation cascade as described above. For small molecules, a cascade of n ~1000 units (total device volume ~1 micron$^3$) completely resolves two mixed molecular species with $D_2/D_1$ = 1.01. As a crude approximation, D ~ 1 / $MW^{1/3}$ for small spherical particles, so this cascade separates small molecules differing by the mass of one hydrogen atom which should be sufficient for most purposes. Structural isomeric forms of the same molecule, such as a-alanine and b-alanine, often have slightly different diffusion coefficients, thus are also easily separable using a diffusion cascade. However, stereoisomeric (chiral) forms cannot be sorted by diffusion through an optically inactive solvent like water.

For large molecules, a 1 million-unit cascade (total device volume ~1000 micron$^3$) provides $D_2/D_1$ ~ 1.00001, sufficient to completely separate large molecules differing by the mass of a single carbon atom. The fidelity of such fine resolutions depends strongly upon the ability to hold constant the temperature of the chamber, since D varies directly with temperature. Device temperature stability will be determined by at least three factors: (1) the accuracy of onboard thermal sensors in measuring T ($\Delta T/T < 10^{-6}$), (2) the rapidity with which the temperature measurement can be taken ($10^{-9}$ to $10^{-6}$ sec), and (3) the time that elapses between the temperature measurement and the end of the diffusive sortation process (which may be of the same order as the gate closing time, ~$10^{-6}$ sec).

Most of the waste heat is generated in this device by forced water sieving. To remain within biocompatible thermogenic limits (~$10^9$ watts/m$^3$), each unit may be cycled once every ~3 millisec, a 0.8% duty cycle of a ~23 microsec sieving stroke. Subject to this restriction, each unit would consume ~1 picowatt in continuous operation. A unit presented with a ~0.1 M concentration of small molecules processes ~$10^6$ molecules/sec (e.g., ~1 gm/hour of glucose using 1 cm$^3$ of n = 1000-unit cascades), or ~$10^4$ molecules/sec for a unit presented with large molecules at ~0.001 M, circulating ~$10^9$ molecules/sec of water as working fluid while running at 340 cycles/sec. Additional chamber segments on each unit, combined with more complex diffusion circuits among the many units in a cascade, should permit the simultaneous complete fractionation of the input feedstock even if hundreds of distinct molecular species are present.

By 1998, diffusion-based separation had been demonstrated in microfluidic devices.

**Nanocentrifugal Sortation**

Nanoscale centrifuges offer yet another method for rapid molecular sortation, by biasing diffusive forces with a strong external field. The well-known effect of gravitational acceleration on spherical particles suspended in a fluid is described by Stokes' Law for Sedimentation:

$$v_t = 2\, g\, R^2 (\rho_{particle} - \rho_{fluid}) / 9\, \eta \qquad \text{...(10)}$$

where $v_t$ is terminal velocity, g is the acceleration of gravity (9.81 m/sec$^2$), R is particle radius, $\rho_{particle}$ and $\rho_{fluid}$ are the particle and fluid densities (kg/m$^3$), and $\eta$ is coefficient of viscosity of the fluid. Particles which are more dense than the suspending liquid tend to fall. Those which are less dense tend to rise ($\rho_{particle} / \rho_{fluid} \sim 0.8$ for lipids, up to $\sim 1.5$ for proteins, and $\sim 1.6$ for carbohydrates in water).

This separation process may be greatly enhanced by rapidly spinning the mixed-molecule sample in a nanocentrifuge device. For ideal solutions (e.g., obeying Raoult's law) at equilibrium:

$$c_2 / c_1 = \exp\left[\left(\frac{MW_{kg}\omega^2}{2R_gT}\right)\left(1 - \frac{\rho_{fluid}}{\rho_{particle}}\right)\left(r_2^2 - r_1^2\right)\right] \qquad \text{...(11)}$$

where $c_2$ is the concentration at distance $r_2$ from the axis of a spinning centrifuge (molecules/m$^3$), $c_1$ is the concentration at distance $r_1$ (nearer the axis), $MW_{kg}$ is the molecular weight of the desired molecule in kg/mole, w is the angular velocity of the vessel (rad/sec), T is temperature (K) and the universal gas constant $R_g$ = 8.31 joule/mole-K. The approximate spinning time $t_s$ required to reach equilibrium is

$$t_s = \ln(r_2 / r_1) / (\omega^2 S_d) \qquad \text{...(12)}$$

where $S_d$ is the sedimentation coefficient, usually given in units of $10^{-13}$ sec or svedbergs. Research ultracentrifuges have reached accelerations of $\sim 10^9$ g's.

Consider a cylindrical diamondoid vessel of density $\rho_{vessel}$ = 3510 kg/m$^3$, radius $r_c$ = 200 nm, height h = 100 nm, and wall thickness $x_{wall}$ = 10 nm, securely attached to an axial drive shaft of radius $r_a$ = 50 nm. A fluid sample containing desired molecules enters the vessel through a hollow conduit in the drive shaft, and the device is rapidly spun. If rim speed $v_r$ = 1000 m/sec (max), then $\omega = v_r / r_c = 5 \times 10^9$ rad/sec ($\omega / 2\pi = 8 \times 10^8$ rev/sec). The maximum

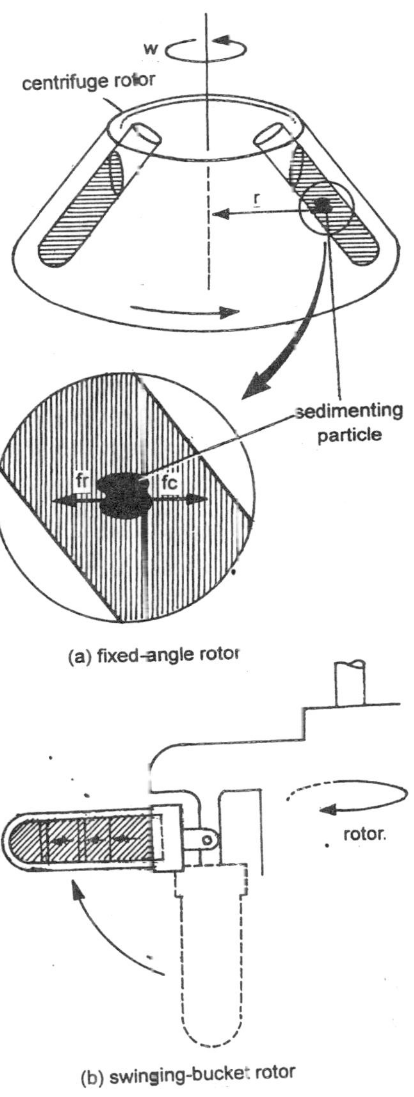

*Fig. 4.2. Centrifuge.*

bursting force $F_b \sim 0.5\ \rho_{vessel}\ v_r^2 = 2 \times 10^9\ N/m^2$, well below the $50 \times 10^9\ N/m^2$ diamondoid tensile strength conservatively assumed by Drexler. Since $S_d$ ranges from $0.1\text{-}200 \times 10^{-13}$ sec for most particles of nanomedical interest, minimum separation time using acceleration $a_r / g = v_r^2 / g\ r_c = 5 \times 10^{11}$ g's, when $r_2 = r_c$ and $r_1 = r_a$, is $t_s = 0.003\text{-}6.0 \times 10^{-6}$ sec. Fluid sample components migrate at $\sim 0.1$ m/sec.

Maximum centrifugation energy per particle $E_c = (MW_{kg} / N_A)\ a_r\ (r_c - r_a) \sim$ 10,000 zJ/molecule, or ~10 zJ/bond for proteins, well below the 180-1800 zJ/bond range for covalent chemical bonds. However, operating the nanocentrifuge at peak speed may disrupt the weakest noncovalent bonds (including hydrophobic, hydrogen, and van der Waals) which range from 4-50 zJ/bond. The nanocentrifuge has mass $\sim 10^{-17}$ kg, requires ~3 picojoules to spin up to speed (bearing drag consumes ~10 nanowatts of power, and fluid drag through the internal plumbing contributes another ~5 nanowatts), completes each separation cycle in $\sim 10^4$ revs ($\sim 10^{-5}$ sec), and processes ~300 micron³/sec which is $\sim 10^{13}$ small molecules/sec (at 1% input concentration) or $\sim 10^9$ large molecules/sec (at 0.1% input concentration).

The nanocentrifuge separates salt from seawater with $c_2/c_1 \sim 300$ across the width of the vessel ($r_c - r_a = 150$ nm); extracting glucose from water at 310 K, $c_2/c_1 \sim 10^5$ over 150 nm. For proteins with $\rho_{particle} \sim 1500$ kg/m$^3$, separation product removal ports may be spaced, say, 10 nm apart along the vessel radius while maintaining $c_2/c_1 \sim 10^3$ between each port. Vacuum isolation of the unit in an isothermal environment and operation in continuous-flow mode could permit exchange of contents while the vessel is still moving, sharply reducing remixing, vibrations, and thermal convection currents between product layers.

Variable gradient density centrifugation may be used to trap molecules of a specific density in a specific zone for subsequent harvesting, allowing recovery of each molecular species from complex mixtures of substances that are close in density. The traditional method is a series of stratified layers of sucrose or cesium chloride solutions that increase in density from the top to the bottom of the tube. A continuous density gradient may also be used, with the density of the suspension fluid calibrated by physical compression. For example, the coefficient of isothermal compressibility $\kappa = (\Delta V_1/V_1) / \Delta P_1 = (\Delta\rho_{fluid} / \rho_{fluid}) / \Delta P_1 = 4.492 \times 10^{-5}$ atm$^{-1}$ for water at 1 atm and 310 K (compressibility is pressure- and temperature-dependent). Applying $P_1 = 12{,}000$ atm to the vessel raises fluid density to 1250 kg/m$^3$, sufficient to partially regulate protein zoning. A multistage cascade may be necessary for complete compositional separation. Protein denaturation between 5000-15,000 atm due to hydrogen bond disruption may limit nanocentrifugation rotational velocity. Protein compressibility may further reduce separability. The balance between the differential densities and the differential compressibilities will determine the equilibrium radius of the protein in the centrifuge; in the limiting case of equal compressibilities for a given target protein and water, there is no stable equilibrium radius.

The nanocentrifuge may also be useful in isotopic separations. For a $D_2O/H_2O$ mixture, $c_2/c_1 = 1.415$ per pass through the device; $c_2/c_1 = 10^6$ is achieved in a 40-unit cascade. Tracer glycine containing one atom of $C^{14}$ is separated from natural glycine using a 113-unit cascade, achieving $c_2/c_1 = 10^6$.

## Membrane Filtration

Filtration through a permeable membrane is closely related to the process of diffusion, since in both cases random molecular motions help carry the process to completion. However, the presence of a

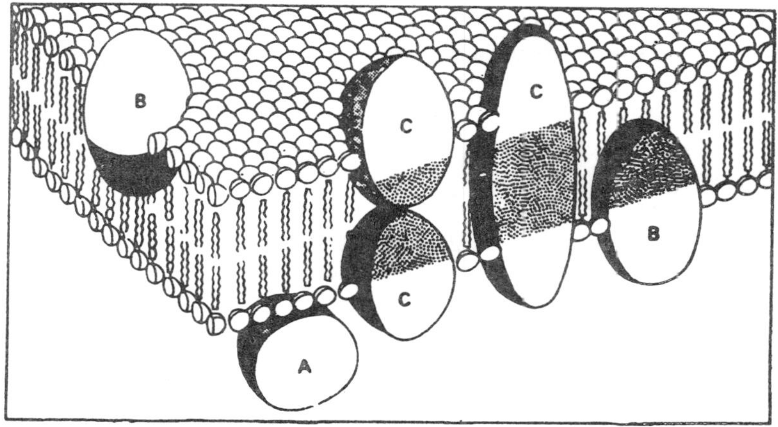


*Fig. 4.3. Plasma membrane.*

membrane adds a new measure of control that is not exploited in simple diffusive transport. This control may be either passive or active, unidirectional or bidirectional, as described below.

**Simple Nanosieving**

Nanometer-scale isoporous molecular sieves (with ovoid, square, or hexagonal holes) are common in almost every taxonomic group of eubacteria and archaeobacteria. Other well-known examples of nanoporous structures are the 6-nm pore arrays found in reverse osmosis and kidney dialysis membranes.

Likewise, it is possible for a nanodevice to sort molecules by simple sieving. In this process, a sample containing particles of various sizes suspended in water passes through a graduated series of filters perforated by progressively smaller holes of fixed size and shape. Between each filtration unit, the filtration residue consists almost exclusively of particles having a narrow range of sizes and shapes. For example, a series of n = 100 filtration units could reliably differentiate an input sample containing particles from 0.2-1.2 nm into 100 separate fractions, each fraction consisting predominantly of particles differing in mean diameter by ~0.01 nm. (In a practical system, several passes would be required to achieve complete discrimination.) A ~0.01 nm difference in molecular diameter corresponds to the mean contribution of ~1 additional carbon atom to the size of a small molecule (MW ~ 100 daltons), or to the mean contribution of ~100 additional carbon atoms to the size of a large molecule (MW ~ 100,000 daltons, ~17,000 atoms). A nanomembrane might even permit the (slow, multi-pass) sieving of oxygen from air, since the molecular diameters of $N_2$ and $O_2$ differ by ~0.01 nm.

Two opposing forces are at work when moving water and solutes through a membrane. One is the osmotic pressure established by the presence of nonpermeating solutes; the other is the hydraulic or fluid pressure. The velocity of material movement depends on the relative values of the osmotic and hydraulic forces, and on the size of the pores in the filter.

Osmotic pressure $p_\pi$ is given by the Donnan-van't Hoff formula:

$$p_\pi = \left(\frac{R_g T(c_2 - c_1)}{MW_{kg}}\right)\left(1 + \frac{Z^2(c_2 - c_1)}{c_1}\right)(N/m^2) \qquad ...(13)$$

where $R_g$ = 8.31 joules/mole-K, T is temperature in kelvins (K), $c_2$ and $c_1$ are solute concentrations on either side of the membrane in kg/m$^3$ ($c_2 > c_1$), $MW_{kg}$ is the molecular weight of the solute in kg/mole, and the $Z^2$ term (dependent upon polymer-polymer interactions) is a correction factor for highly concentrated solutions which for some solvents and temperatures may equal zero; Z = net solute charge number and $c_s$ is the concentration in kg/m$^3$ of a second solute, as for example when the first solute is a protein and the second solute is salt, as in human serum. Water at 310 K dissolves a maximum of $c_2$ = 370 kg/m$^3$ of sodium chloride (a 37.0% solution, by weight), and $MW_{kg}$ = 0.05844 kg/mole for NaCl, so for salt water solutions the theoretical maximum $p_\pi \sim 1.6 \times 10^7$ N/m$^2$ ~ 160 atm. Natural bloodstream concentrations of salt produce $p_\pi$ ~ 3 atm. Since osmotic pressure depends on the number of molecules present, the contribution from large molecules is usually negligible. For instance, total protein concentration in human blood serum is $c_2$ ~ 73 kg/m$^3$, $MW_{kg}$ ~ 50 kg/mole (~50,000 daltons), so $p_\pi$ ~ 0.04 atm.

In theory, extremely large hydraulic counterforces up to $10^5$ atm may be applied in nanomechanical systems, for example by a piston, to overcome osmotic backpressure. As a practical matter, however, rapidly pushing small molecules at high pressure through nanoscale holes is an effective method for generating significant amounts of waste heat; a design compromise is required.

Consider a simple sorting apparatus in which a square piston is used to compress solvent fluid (say, water) trapped in a chamber $h_{chamber}$ in length and $L_{chamber}^2$ in cross-sectional area, forcing the fluid to filter through pores of radius $r_{pore}$ (~ target molecule radius) covering a fraction $\alpha_H$ (~50%) of the surface of a square nanoscale sieve of thickness $h_{sieve}$ and area $L_{sieve}^2$. Solute which is dissolved or suspended in the solvent is sorted based on molecular size; a sequence of sieving runs using sieves having progressively smaller pore radii produces an ordered sequence of molecular size fractions. (In small molecules, adding the mass of one hydrogen atom increases the mean linear dimension of the molecule by 0.11%.)

The first design constraint on this system relates to its maximum operating pressure. If $\Delta P$ is applied pressure (N/m$^2$), then avoiding boiling the solvent water and denaturing proteins requires at least that $\Delta P_{max} < C_V \Delta T_{boil}$, where the heat capacity of water $C_V = 4.19 \times 10^6$ joules/m$^3$-K and $\Delta T_{boil} = 373$ K - 310 K = 63 K give $\Delta P_{max} <$ 2600 atm. The designs presented below operate at 6% of this maximum ($\Delta T \sim 3$ K) or less. There are two major design constraints on the duration of the power stroke, or $t_p$:

***Molecular rotation constraint***

Flow through the sieve must be slow enough to allow molecules to align with the holes. Assuming round pores, the total number of pores in the sieve is $N_{pore} = \alpha_H L_{sieve}^2/\pi r^2$. The volume processing rate (m$^3$/sec) of the sieve $`V_{sieve} = `V_{chamber} = h_{chamber} L_{chamber}^2/t_p$, hence the molecule processing rate is $`N = `V_{chamber} c_{target}$ (molecules/sec) where $c_{target}$ is the concentration of target molecules (molecules/m$^3$). At any one time, each pore channel through the sieve can hold at most $N_{channel} = h_{sieve}/2 r_{pore}$ molecules in single file; during each power stroke, at most $N_{stack} = `N t_p/N_{pore}$ molecules pass through each pore. Hence the time available for molecular rotation $t_{rot} = t_p (N_{channel}/N_{stack})$, which assumes the layer of rotating target molecules in the vicinity of the pores approximates sieve thickness $h_{sieve}$, a reasonable assumption as long as the typical molecular diffusion time (across a distance $h_{sieve}$) $<< t_{rot}$. Taking $N_{rot} \sim 10$ as the mean number of molecular revolutions needed to ensure proper pore alignment with noncircular sieve holes (the most difficult case), then, from Eqn. 2, $\Delta\alpha = (kT\, t_{rot} / 4 \pi \eta r_{pore}^3)^{1/2} \geq 2 \pi N_{rot}$, where $\eta$ is solvent viscosity ($1.1 \times 10^{-3}$ kg/m-sec for plasma) at T = 310 K. Solving for minimum $t_p$ gives:

$$t_p \geq \frac{32\pi^4 N_{rot}^2 \eta c_{target} h_{chamber} r_{pore}}{kT \alpha_H h_{sieve}} \text{ (sec)} \qquad \text{...(14)}$$

***Pressure/flow constraint***

Flow through the sieve must be fast enough to establish a sufficient pressure to oppose osmotic backflows. From the Hagen-Poiseuille law, the volume processing rate through each pore is $`V_{pore} = \pi r_{pore}^4$

$\Delta P_{sieve}$ / 8 $\eta$ $h_{sieve}$ and the volume processing rate through the entire sieve is $'V_{sieve} = N_{pore}\ 'V_{pore} = 'V_{chamber}$; solving for maximum $t_p$ gives:

$$t_p \le \frac{8\eta\, h_{chamber}\, h_{sieve}\, L^2_{chamber}}{\alpha_H\, \Delta P_{sieve}\, r_{pore}\, L^2_{sieve}} \text{(sec)} \qquad ...(15)$$

Equating these two bracketing constraints, $h_{sieve} \ge 150$ nm for large molecules ($r_{pore}$ ~ 5 nm) but $h_{sieve} \ge 1$ nm for small molecules ($r_{pore}$ ~ 0.32 nm).

As a final constraint, power released by fluid flow through the chamber and sieve must not exceed safe thermogenic limits. Given the maximum safe power density for *in vivo* nanomachines as $D_n = 10^9$ watts/m$^3$, then:

$$D_{device} = \frac{P_{device}}{(h_{chamber} + h_{sieve})\, L^2_{chamber}} \le D_n \qquad ...(16)$$

where $D_{device}$ = device power density (watts/m$^3$), total device power $P_{device} = P_{chamber} + P_{sieve}$ (watts), chamber fluid flow power $P_{chamber}$ = $\pi\ L_{sieve}^4\ \Delta P_{chamber}^2$ / 128 $h_{chamber}$ $\eta$, sieve fluid flow power $P_{sieve} = \alpha_H$ $L_{sieve}^2\ r_{pore}^2\ \Delta P_{sieve}^2$ / 8 $h_{sieve}$ $\eta$, and $\Delta P_{chamber}$ = 16 $\alpha_H$ $h_{chamber}$ $r_{pore}^2$ $\Delta P_{sieve}$ / $\pi$ $L_{sieve}^2$ $h_{sieve}$. For small molecules such as NaCl or glucose, $\Delta P_{sieve} \ge 160$ atm to overcome maximum osmotic backpressure; for large molecules (e.g., ~50,000 dalton proteins), we assume $\Delta P_{sieve} \ge$ 1 atm to ensure sieving. The following designs are not optimized but illustrate the tradeoffs involved.

For small molecules ($r_{pore}$ ~ 0.32 nm), an exemplar ~1 micron$^3$ device has $h_{chamber}$ = 1 micron, $L_{chamber} = L_{sieve}$ = 0.6 micron, $h_{sieve}$ = 1.5 microns, $t_p$ = 0.016 sec, $\Delta P_{sieve}$ = 160 atm, $\Delta P_{chamber}$ = 0.0001 atm. Piston velocity ~ 60 micron/sec and 'V = 2.5 $\times$ $10^{-17}$ m$^3$/sec for a 0.1 M solution of target molecules, yielding a processing rate of 1.5 $\times$ $10^9$ molecules/sec (1.5 $\times$ $10^{-16}$ kg/sec); the device processes its own mass every ~7 sec or every ~430 power strokes. Device power $P_{device}$ = 400 pW and power density $D_{device}$ = 4 $\times$ $10^8$ watts/m$^3$.

For large molecules ($r_{pore}$ ~ 5.0 nm), an exemplar ~1 micron$^3$ device has $h_{chamber}$ = 1 micron, $L_{chamber} = L_{sieve}$ = 0.9 micron, $h_{sieve}$ = 0.15 microns, $t_p$ = 0.010 sec, $\Delta P_{sieve}$ = 1 atm, $\Delta P_{chamber}$ = 0.0004 atm. Piston velocity ~ 100 micron/sec and 'V = 9.8 $\times$ $10^{-17}$ m$^3$/sec for a 0.001 M solution of target molecules, yielding a processing rate of 5.9 $\times$ $10^7$ molecules/sec (4.9 $\times$ $10^{-15}$ kg/sec); the device processes its own mass every ~0.2 sec or every ~20 power strokes. Device power $P_{device}$ = 80 pW and power density $D_{device}$ = 8 $\times$ $10^7$ watts/m$^3$.

Sieve pores can become clogged by particles of radius $R \sim r_{pore}$ if the applied hydraulic pressure $\Delta P_{clog}$ exceeds the thermal energy of the trapped particles, or:

$$\Delta P_{c\,log} > 9kT / 8\pi R^3 \ (N / m^2) \qquad ...(17)$$

By this criterion, $\Delta P_{clog} > 500$ atm for small molecules and $\Delta P_{clog} > 0.1$ atm for large molecules. From the values of $\Delta P_{sieve}$ given above, clogging is unlikely for small molecules but is possible at the highest concentrations of large molecules. In 310 K water, large molecules diffuse ~3 nm and small molecules diffuse ~17 nm in ~$10^{-7}$ sec, just far enough to clear the hole, so a ~10 MHz sawtooth pressure profile imposed on the power stroke may ensure sufficient backflushing action to avoid serious blockages. To reduce the possibility of clogging due to surface force adhesion, as a design criterion the work of adhesion should be reduced to $W_{adhesion} < \Delta P_{sieve} \, r_{pore} \sim 5 \times 10^{-3}$ J/m$^2$ for small molecules and ~ $0.5 \times 10^{-3}$ J/m$^2$ for large molecules likely to come into contact with sieve pore surfaces. Clogging due to long-term random polymerizations can be minimized by periodically exchanging the entire contents of the input chamber with fresh solution, by operating the device at reduced power density, or by periodically replacing the sieve.

**Dynamic Pore Sizing**

A more efficient nanosieve system can be designed if pore size and shape can be actively modified during device operation, as for example by exchanging filters (from a membrane library stocking various pore sizes) each half cycle. Better, if pores can be reliably dilated or constricted in place during a period of time $\Delta t << t_p$, then filtration cascades can be more rapidly reconfigured to match changing input feedstock characteristics or to extract varying selections of desired molecules at will. Additionally, fully differentiating sieving cascades can be collapsed into a single unit, providing more compact devices especially useful in chemical sensor systems requiring preconcentration of sample. Control of pore shape should also provide finer discrimination among molecules of similar size but different shape, such as some isomers of nonchiral molecules.

Two or more overlapping surfaces containing regular arrays of perforations of fixed size and shape can conveniently generate a wide variety of pore geometries. Control of pore geometry is achieved by sliding or rotating one surface relative to the other surface by a small increment.

Circular dilating apertures can also be constructed using a matched set of overlapping segments, which may be driven either radially or tangentially to enlarge or contract the hole like an irising camera diaphragm, consistent with Akey's model of the nuclear pore complex present on the cell nucleus surface. Diaphragming mechanisms may be vertically staggered to maximize areal hole density in filtration surfaces (at the cost of increased vertical rugosity). Filters constructed of hydrogen-passivated diamondoid can have pores with <0.1 nm feature sizes, although H-free fullerene materials might avoid any possibility of dehydrogenation shearing.

Methods of positioning surfaces to accuracies of ~0.01 atomic diameter (!0.001 nm) have already been discussed. Assuming pore sizing blades require ~25 $nm^2$ of diamondoid contact surface per pore and each blade travels 25 nm at 0.01 meter/sec during one cycle, sliding friction dissipates ~0.01 zJ/pore, or ~4 × $10^{-18}$ watts/pore during each 2.5 microsec resizing cycle. Since fluid friction approaches kT for nanometer-size holes changing size in ~$10^{-9}$ sec, maximum blade speed is ~1 m/sec and the fastest resizing cycle is ~$10^{-8}$ sec.

A single sieving unit with controllable pores can be moderately efficient. Consider a design similar to that already described, except for a separate chamber and piston on either side of the filter block. Suppose that the particles desired to be extracted are of radius r, and the next smallest possible pore size is r - Δr. The device operates in two phases. In the first phase, the sample is placed in the first chamber, pore size is set equal to r, then the first piston forces the fluid through the membrane. Particles larger than r remain behind and are flushed from the first chamber. The pores then contract to r - Δr, and the second piston pushes the remaining filtrate back into the first chamber. After this second phase, particles of radius ~(r ± Δr) remain in the second chamber at significantly higher concentration and may be removed for further use. Analogous double-sieve editing paradigms are commonplace in biological systems.

Other designs might work equally well, such as a 3-chamber flowthrough design using a variable pore membrane with pores of size r between the first and second chamber, a membrane with pores of size r - Δr between the second and third chamber, and a piston at either end (one pushing, one pulling), thus concentrating molecules of size r ± Δr in the central chamber. M. Krummenacker suggests fixed chambers with a moving sieve operated as a dragnet. Filtration processes may be most useful in performing complete separations of complex mixtures. But they are inefficient in the sense that the energy expended to orient molecules passing through pores is wasted if the molecules are allowed to randomize on the other side; a eutactic mill-like molecule handling system might preserve this order and greatly improve energy efficiency.

**Gated Channels**

Besides controlling nanopore size and shape, individual molecular transport channels can be gated either mechanically (e.g., ligand gating) or electrically (e.g., voltage gating). Either method might usefully be employed to control molecular transport through the surfaces of medical nanodevices in a process that could very loosely be described as molecular transistor gating.

A good example of mechanical gating in biology is the nicotinic acetylcholine receptor channel, probably the best understood ligand-gated channel. Nerve impulses are communicated across neuromuscular junctions and autonomic ganglia via neurotransmitters such as acetylcholine. STM images confirm that the receptor itself is cylindrical, a bundle of 5 rod-shaped polypeptide subunits arranged like barrel staves with outside diameter ~6.5 nm. The receptor protrudes 6 nm on the synaptic side of the membrane and 2 nm on the cytoplasmic side. The water-filled channel pore lies along the symmetry axis, lined by 5 α-helices, with a 2.2 nm wide mouth on the synaptic surface, a 0.65 nm waist where the structure dives through the cell membrane, and a 2 nm wide cytosolic exit.

Normally, the channel is closed and no ions may pass. In this closed state, the channel is occluded at the waist by a ridge of large residues forming a tight hydrophobic ring. Each subunit has a bulky leucine at the bend in the α-helix, a critical position. When two acetylcholine molecules bind to the receptor, these helices allosterically tilt, shifting the position of the ridges. The pore becomes open because it is now lined with small polar residues rather than by large hydrophobic ones. This conformational change allows $2.5 \times 10^7$ $Na^+$ ions/sec to flow through the channel, about 10% of the diffusion-limited rate. (Anions like $Cl^-$ cannot enter the pore because they are repelled by rings of negatively charged residues positioned at either end of the receptor.)

Acetylcholine binding opens the gate in less than 100 microsec under physiologic conditions. Subsequent rapid destruction of acetylcholine by acetylcholinesterase, an enzyme tethered to the membrane surface by a covalently attached glycolipid group, closes

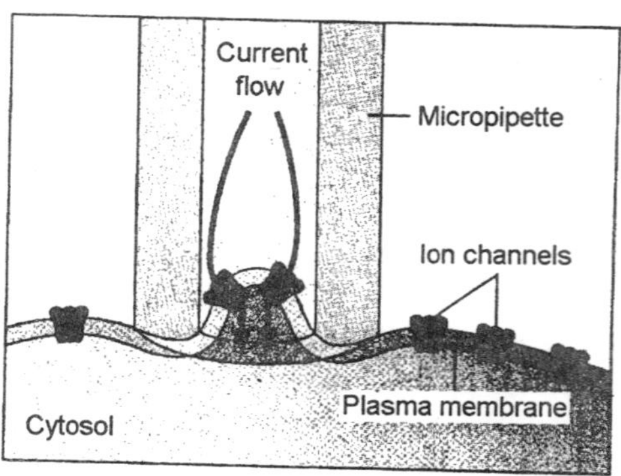


*Fig. 4.4. Ion channels*

the gate in ~1 millisec. Much faster gating action (~$10^{-8}$ - $10^{-6}$ sec) could be achieved by nanodevices operating variable-scale nanopores in response to sensor data or other control signals. Such signals could drive the insertion or retraction of diamondoid rods, wedges, or trapdoors across the channel lumen to regulate the transmission of molecules having specific sizes, shapes, and charge distributions.

Transport channels through nanodevice surfaces may also be gated electrically. In contrast to the acetylcholine receptor, which is relatively nondiscriminating and allows both inorganic and organic cations to pass, the voltage-gated calcium channel has a highly discriminating mechanism with a $Ca^{++}$:$Na^{+}$ permeation ratio on the order of 1000:1. (The high specificity of the voltage-gated $Ca^{++}$ channel is a consequence of a single-file pore mechanism involving a pair of specific $Ca^{++}$ ion binding sites. Selectivity is assured if either of the two sites is occupied by $Ca^{++}$, as monovalent ions do not bind strongly enough to the free site or generate sufficient electrostatic repulsion to push the first $Ca^{++}$ ion through the channel.) Potassium channels are 100 times more permeable to $K^{+}$ than to $Na^{+}$, and sodium channels favour the passage of $Na^{+}$ over $K^{+}$ by a factor of 12. All three of these voltage-gated channels are important in the generation and conduction of neural action potentials.

A nerve impulse is an electrical signal produced by the flow of ions across the plasma membrane of a neuron. Neuron interiors have high concentrations of $K^{+}$ and low concentrations of $Na^{+}$. The resting potential of a neuron is 60 mV. An action potential may be generated when the membrane potential is slightly depolarized to 40 mV. This opens the $Na^{+}$ voltage-gated channels, rapidly accelerating depolarization to a peak of +30 mV in ~1 millisec. Then $Na^{+}$

channels close and $K^+$ channels open, allowing $K^+$ ions to exit the cell, restoring the 60 mV resting potential. Only ~1 ion of every ~$10^6$ $Na^+$ and $K^+$ ions present in the local extracellular medium and the axoplasm participate in each such nerve impulse.

The sodium channel is a single polypeptide chain with four repeating units. Each repeating unit folds into six transmembrane alpha helices, including one that is positively charged called the S4 helix. The S4 helix is the voltage sensor that triggers the opening of the gate. Three positively charged residues on each S4 helix are paired at the resting membrane potential with negative charges on other transmembrane helices in a staircase geometry. The initial small depolarization event produces a spiral motion of each S4 accompanied by the net movement of one or two charges to the extracellular side of the membrane, essentially turning this left-handed hydrogen-bonded "molecular screw" through a ~60° rotation. This outward 0.5-nm translation of the four S4 segments opens the sodium gate by removing a steric barrier to ion flow. The energy cost of moving ~6 electrical charges (~$10^{-18}$ coul) from the cytosolic to the extracellular side of the membrane against a ~100 mV potential (thus opening the gate in ~75 microsec) is ~100 zJ. Quantum tunneling activation of sodium channels, taking 1-1000 microsec, has been analyzed by Chancey.

Artificial ion-gated polymer membranes were reported in 1982, protein engineering of switchable pore-forming proteins is well-known, and "intelligent gels" are being developed that can change size and molecular porosity in response to chemical, electrical or thermal stimuli. In 1998, however, electroporation was a more commonly used method in biological research and a useful technique for "transfecting" cells in genetic studies. Electroporation employs a brief intense pulse of electricity to provide a force that opens cellular pores, enabling the insertion of macromolecules like DNA into cells of interest; laser pulses reduce cell loss to 10% by using a square-wave pulse to effect rapid and reversible pore formation. Artificial pH-gated 200-nm diameter nanopores and natural pH-gating of virion pores in the cowpea chlorotic mottle virus were demonstrated in 1998.

The first true voltage-gated nanomembrane was fabricated by Charles Martin and colleagues in 1995. This membrane consists of cylindrical gold nanotubules with inside diameters as small as 1.6 nm. When the tubules are positively charged, cations are excluded and only negative ions are transported through the membrane. When the membrane receives a negative voltage, only positive ions are transported through the tubules. Nanodevices may combine voltage gating with pore size and electrosteric constraints to achieve precision transport control with moderate molecular specificity at diffusion-limited throughput rates. In 1997, an ion channel switch biosensor with sub-picomolar sensitivity and quantitative detection time of ~600 sec was demonstrated by an Australian research group.

## Receptor-Based Transport

The most efficient of all modes of molecular transport involves receptor sites capable of recognizing and selectively binding specific molecular species. Many receptors reliably bind only a single molecular type; others, such as sugar transporters, can recognize and transport several related sugar molecule types. In nanomechanical systems, artificial binding sites of almost arbitrary size, shape, and electronic charge may be created and employed in the construction of a variety of highly efficient molecular sortation and transport devices, described below.

### Transporter Pumps

While gated channels enable ions to flow rapidly through membranes in a thermodynamically downhill direction, active pumps use a source of free energy to force an uphill transport of ions or molecules. In biology the energy supply is usually ATP or photons of light; medical nanomachines can make use of vastly more diverse energy sources. Actually, the term "pump" may be something of a misnomer because the action is highly specific—only one or a very small number of molecular species are selectively transported.

Molecular pumps generally operate in a four-phase sequence: (1) recognition (and binding) by the transporter of the target molecule from a variety of molecules presented to the pump in the input substrate; (2) translocation of the target molecule through the membrane, inside the transporter mechanism; (3) release of the molecule by the transporter mechanism; and (4) return of the transporter to its original condition, so that it is ready to accept another target molecule. Such molecular transporters that rely on protein conformational changes are ubiquitous in biological systems.

The minimum energy required to pump molecules is the change in free energy $\Delta G$ in transporting the species from one environment having concentration $c_1$ to a second environment having concentration $c_2$, given by:

$$\Delta G = kT \ln(c_2 / c_1) + \frac{Z_e F \Delta V}{N_A} \quad \ldots(18)$$

where k = 0.01381 zJ/K (*Boltzmann constant*), T = 310 K, $Z_e$ is the number of charges per molecule transported (i.e., the valency), F = $9.65 \times 10^4$ coul/mole (*Faraday constant*), $\Delta V$ is the potential in volts across the membrane, and $N_A$ is Avogadro's number. So for example, transport of an uncharged molecule across a $c_2/c_1 = 10^3$ gradient (typical in biology) costs ~30 zJ. An extremely aggressive $c_2/c_1 = 10^6$ concentration gradient costs ~60 zJ/molecule, plus another ~30 zJ/ion if we are moving $Ca^{++}$ ions against a 100 mV potential. An artificial nanopump of dimension ~10 nm moving at a conservative ~1 cm/sec velocity operates at MHz frequencies, transporting ~$10^6$ molecules/sec for a continuous power consumption of ~0.03 pW at $c_2/c_1 = 10^3$. Such a pump has a mass ~$10^{-21}$ kg.

Transporter pumps need not be limited to the movement of a single molecular species in a single direction, which biochemists call a uniport transport mechanism. Numerous well-known biological systems are capable of moving two molecules simultaneously in one direction (*symport mechanisms*), two molecules sequentially in opposite directions (*antiport mechanisms*), and charged molecules in one direction only, thus building up an electrical charge on one side of the membrane (electrogenic mechanisms). Such pumps exist in nature for numerous ions, amino acids, sugars, and other small biomolecules. Active drug efflux systems and multidrug resistance are made possible by the

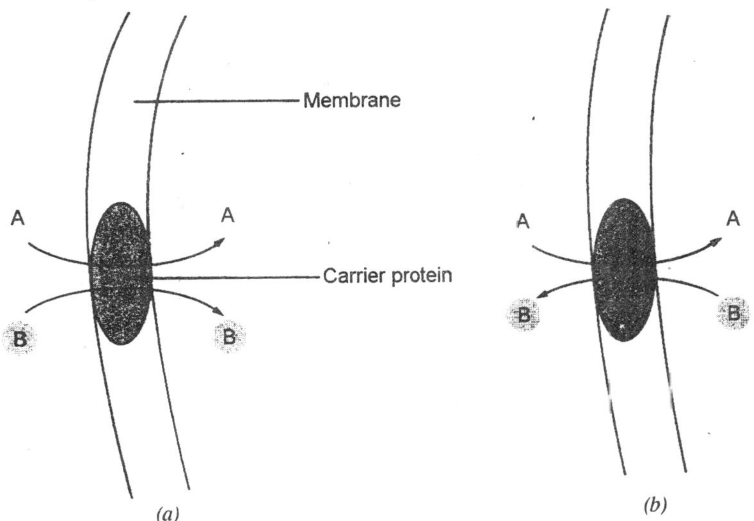


*Fig. 4.5. Showing (a) symport; (b) antiport*

expression of bacterial genes coding for molecular pumps that are constantly evolving new specificities to increasing numbers of microbicidal drugs.

Three $Na^+$ and two $K^+$ ions are transported per 10 millisec cycle, requiring the hydrolysis of one ATP molecule to ADP to drive the conformational changes. (More than one-third of the ATP consumed by a resting animal is used to pump these two ions.) Hydrolysis of ATP liberates ~80 zJ/molecule of free energy, so the antiporter is transporting $Na^+$ and $K^+$ at a cost of 16 zJ/ion at a 0.5 KHz frequency. Pump site density is ~1000/micron$^2$ of cell membrane in neural C fibers. (The $Na^+$ - $K^+$ pump can also be operated in reverse to synthesize ATP from ADP by exposing the mechanism to steep ionic gradients.) By contrast, artificial nanomechanical antiporter and symporter devices will operate at MHz frequencies. They should be able to transport much larger molecules, and may also be fully reversible.

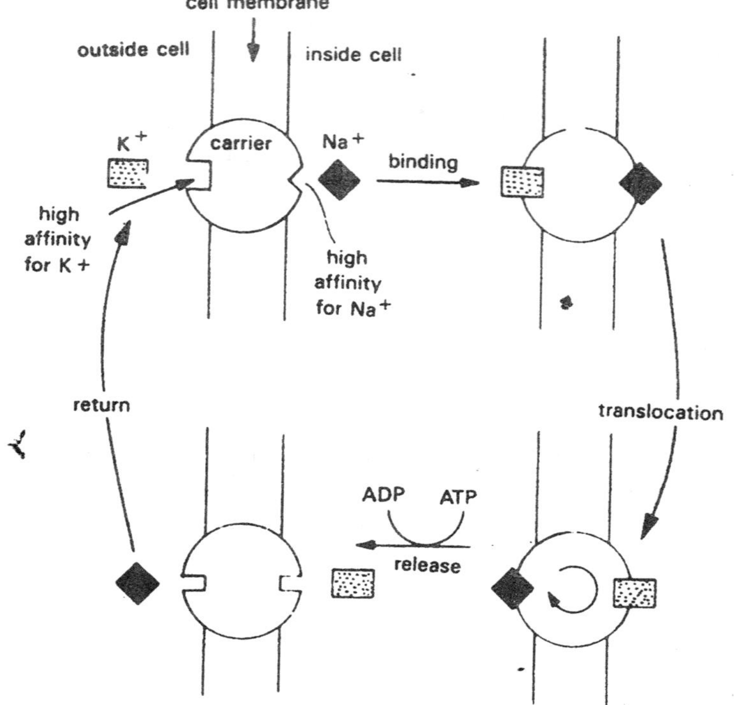


*Fig. 4.6. $Na^+$ and $K^+$ pump.*

**Sorting Rotors**

Drexler's molecular sorting rotor is a related class of nanomechanical device capable of selectively binding molecules from solution and then transporting these bound molecules against concentration gradients. The archetypal sorting rotor is a disk with 12 binding site "pockets" along the rim exposed alternately to the external solution and interior chamber by axial rotation of the disk. (Other designs may have more, or fewer, pockets.) Each pocket selectively binds a specific molecule when exposed to the solution. Once the binding site rotates to expose it to the interior chamber, the bound molecules are forcibly ejected by rods thrust outward by the cam surface (or using some other means by which receptor affinity can be adjusted during the inbound transport process). In the case of protein molecules, the debinding geometry must be carefully designed to avoid denaturation during ejection. Also, the rotor implicitly assumes that target molecules remain in the liquid or gaseous state after importation. M. Krummenacker observes that most bloodborne molecular species will precipitate as solids unless they are well-solvated; thus nanomedical sorting systems may require internal solvent or in some cases should be made completely eutactic. The discovery of positionally disordered water molecules resident inside protein hydrophobic cavities suggests that good rotor designs may also need to include solvent drainage channels.

Molecular sorting rotors can be designed from about $10^5$ atoms (including housing and pro rata share of the drive system), measuring roughly 7 nm × 14 nm × 14 nm in size with a mass of $2 \times 10^{-21}$ kg. Rotors turn at ~86,000 rev/sec with a conservative rim speed of 2.7 mm/sec and an almost negligible drag power of $\sim 10^{-16}$ watts against the fluid, sorting small molecules at a rate of $10^6$ molecules/sec with laminar flow. From Eqn. 18, the energy cost of small-molecule sortation at 310 K ranges from ~10 zJ/molecule at low pressures ($c_2/c_1 = 10$) up to ~40 zJ/molecule when pumping against the highest head pressures ($c_2/c_1 = 10^4$, ~30,000 atm for natural bloodstream concentrations of salt with osmotic $p_p \sim 3$ atm), consuming 0.01-0.04 pW per device in continuous operation. Rotors are fully reversible, so they can be used to load or unload target molecules depending on the direction of rotor rotation. Cylindrical rotors with many receptor rows are somewhat more energy-efficient, and rotor lifetimes should be $> 10^6$ sec.

Typical molecular concentrations in the blood for target molecules of nanomedical interest are $\sim 10^{-11}$ - $10^{-3}$ molecules/nm$^3$, which should be sufficient to ensure ~99% occupancy of rotor binding sites. Rotors targeting serum hormones and other low-concentration species at the parts-per-billion level must slow to <1 rev/sec to ensure complete receptor occupancy and to avoid exceeding diffusion limits.

Sorting of ionically charged species can use binding sites that display opposite charge, effectively increasing the affinity of the binding site for the charged species. Many ionic species dissolved in water are actually more complex than their symbolic representation would suggest. As an example, a naked proton ($H^+$) in water is always highly hydrated, usually as $H_5O_2^+$ or $H_7O_3^+$, or even as $H_9O_4^+$ in strong acid solutions. Small ions from $Li^+$ to $I^-$ are also found in solvent cages, bound with energies $\geq$330 zJ relative to vacuum; for instance, $Li^+$ and $I^-$ are coordinated to 46 water molecules. As a result, the design of an appropriate binding site or filtration process for such ions can likewise become more complex. The existence of biological channels that show remarkable selectivity for specific ions (e.g., $Na^+$ channels that largely exclude $K^+$ ions, and vice versa) provides one design approach for dealing with such species.

Drexler has also proposed cascades of sorting rotors to achieve high fidelity purification and a contaminant fraction of $<10^{-15}$. However, since only $\sim 10^{10}$ small molecules can be stored in 1 micron$^3$ volume (typical for nanomedical devices), 100% process purity requires a contaminant fraction of only $<10^{-10}$ which can be ensured in micron-scale nanosystems using at most 5 stages, starting from a dilute input substrate containing only 1 part per billion of the target molecule with each stage providing a concentration factor of $\sim 10^4$. For statistically pure extractions of more common molecules in blood and cytoplasm, a 3- or 4-stage cascade will usually suffice. Note that the optimal receptor structure may differ at different stages in a cascade, and that each 12-arm outbound rotor can contain binding sites for 12 different impurity molecules. The first-stage receptor will likely pass only a relatively small number of different contaminant species, so the number of outbound rotors in the entire system can probably be reduced to a small.

**Internal Transport Streams**

After pump mechanisms described above have reliably sorted externally-encountered target molecules into reservoirs filled with species of a single type, a medical nanodevice may require these molecules to be transported to specific internal locations for further processing. Bulk fluid flow or fluidized (solvated or suspended) transport through nanopipes may suffice for some purposes. However, in many cases it will be necessary to present reagent molecules to other subsystems as a well-ordered stream of precisely positioned moieties transported *in vacuo*, especially for mechanochemical operations.

For this purpose, Drexler proposes molecular mills—eutactic systems of nanoscale belts moving over rollers, with reagent-binding devices mounted on the belt surface. This class of device can be assembled into complex molecular transportation networks using conditional switching, crossed-axis belting, and transit speed/frequency multipliers, and may also be employed to drive mechanosynthetic chemical reactions. The benchmark mechanism uses 10-nm diameter rollers to carry closely packed reagent devices measuring 4 nm $\times$ 4 nm $\times$ 2 nm, or 32 $nm^3$. A 20-roller mill mechanism 1 micron long has a ~2 micron long belt with 500 reagent devices and delivers $10^6$ molecules/sec at a belt speed of 4 mm/sec. Total power dissipation is ~$1.4 \times 10^{-18}$ watts, a rate of ~0.001 zJ per moiety (or per reagent device) delivered or ~$10^{-6}$ zJ/nm traveled per reagent device. Total mill mechanism mass is ~$6 \times 10^{-20}$ kg.

An alternative to roller/belt mill mechanisms is a nonconnected stream of pallets pushed along tracks, also *in vacuo*. Such tracks may include merging junctions, distribution junctions, multiplane crossings and switching stations, as well as straight and curved sections. Assuming each pallet is a 32 $nm^3$ reagent device held to the track by pins in grooves resembling cam followers, energy dissipation by phonon scattering is given approximately by:

$$P_{drag} = \frac{4}{3}\frac{\varepsilon_p\, \omega_{therm} v^2}{v_{sound}} \quad \ldots(19)$$

where $\varepsilon_p$ = $2 \times 10^8$ joules/$m^3$ (phonon energy density), $s_{therm}$ (a thermally-weighted scattering cross section) ~$10^{-20}$ $m^2$ for reagent devices of mass m = $10^{-22}$ kg assuming a sliding contact of stiffness ~30 N/m in a moderately stiff medium, v = 4 mm/sec sliding speed, and $v_{sound}$ = $10^4$ m/sec (~speed of sound in diamond), giving $P_{drag}$ ~$4 \times 10^{-21}$ watts per reagent device, or $P_{drag}$ / v ~ $10^{-6}$ zJ/nm traveled per reagent device (pallet). Note that volume containerization of pallet-transported molecules is least efficient at the smallest scales, where surface area per unit enclosed volume is highest, since energy usage is proportional to the surface area of the carrier. Containerization of n >> 1 molecules for large-pallet transport is more efficient.

A less energy-efficient, but far more versatile, internal molecular transport device is the 100-nm telescoping manipulator arm has already been described. This flexible $\sim 10^{-19}$ kg device employs a binding tip to pick and place small and large molecules alike, moving them at $\sim 1$ cm/sec with repeatable placement accuracy of 0.04 nm. Multiple devices can be used to establish an internal ciliary transport system; standardized volume containerization of molecules permits rapid stereotypical handoff motions and efficient parcel routing. Conveyance through a 100-nm arc takes $10^{-5}$ sec consuming 0.1 pW while the arm is in motion, or $\sim 10$ zJ/nm traveled per reagent molecule or per container transported (vs. $\sim 1000$ zJ per typical covalent bond).

In molecular cytobiology, vesicles and organelles are transported throughout the interior of a cell by riding on microtubular cables crisscrossing the cytosol. For example, neural vesicles show transport speeds up to 2-4 microns/sec.

## Molecular Receptor Engineering

Molecular recognition requires a detailed surface complementarity between the target molecule and its receptor. The interplay of various molecular forces between ligand and receptor causes them to selectively bind together, typically engineered to occur in $\sim 10^{-6}$ sec. It is useful first to briefly review and quantify the principal physical forces at work.

### Physical Forces in Molecular Recognition

Covalent bonds, which occur when atoms share electrons, are the strongest bonds. Aside from metals and salts, most material objects are made of atoms held together by covalent bonds. The atoms comprising biological molecules like proteins, nucleic acids and lipids are strung together mostly by single, double, or triple covalent bonds, as are receptors and diamondoid nanomechanical structures. Interatomic bond strengths range from 181 zJ/bond for O-F up to 1785 zJ/bond for C-O. Covalent bond lengths range from 0.10-0.16 nm within CHON-atom molecules, giving a typical covalent bond rupture force of $\sim 10$ nN/bond.

But as Jean-Marie Lehn points out, "there is a chemistry beyond the molecule" —noncovalent supramolecular chemistry. The bonds employed in molecular recognition are weak noncovalent bonds. Noncovalent bonds are largely responsible for the secondary and higher order structure of macromolecules. On a per-bond basis, noncovalent bonds are 1-3 orders of magnitude weaker than covalent bonds. However, the possibility of combining within a limited area a great number of noncovalent bonds having complementary elements allows the formation of a large specific association whose affinity may be of the same order of magnitude as a covalent bond. The high combinatorial diversity provided by many complementary elements allows numerous orthogonal specific associations, enabling self-assembly of many components; by comparison, covalent chemistry offers a poor diversity of reactivities. An additional advantage is that the formation of noncovalent bonds often is not hindered by high energy barriers. At least five types of noncovalent bonds may be distinguished: electrostatic, hydrogen, van der Waals, π aromatic, and hydrophobic.

***Electrostatic bond***

The electrostatic bond between two charged particles (e.g., the "salt bridge" in proteins) is a dipole interaction whose energy $E_e$ is given by Coulomb's law as:

$$E_e = \left(\frac{e^2}{4\pi\,\varepsilon_0}\right)\left(\frac{Z_1 Z_2}{\kappa_e r}\right)\exp(-K_{dh} r) \text{ (joules)} \qquad ...(20)$$

where e = $1.60 \times 10^{-19}$ coul (elementary charge), $\varepsilon_0 = 8.85 \times 10^{-12}$ farad/m (permittivity constant), $Z_1$ and $Z_2$ are the numbers of attractant charges, r is the distance between the charges, and $\kappa_e$ is the dielectric constant (74.3 for pure water at 310 K, usually reduced to ~40 in a hydrophobic environment). The bond is strengthened if the charges are in a hydrophobic environment. Conversely, the presence of electrolytes weakens the bond energy due to a shielding effect, given by $K_{dh}$, the Debye-Huckel reciprocal length parameter, which has a value of 1.25 $nm^{-1}$ for 0.15 M NaCl (~1% solution, ~human blood). Thus two unit charges separated by 0.3 nm produce an interaction energy of $E_e$ = 19 zJ in a hydrophobic environment, 10 zJ in pure water, and 6.3 zJ in 1% salt water. Long-range electrostatic trapping has been observed in single-protein molecules at liquid-solid interfaces, raising the implication that the interaction of protein molecules with biological cell surfaces may be much more efficient than predicted by random diffusion.

Most isolated amino acids in neutral solution are zwitterionic—the molecule has no overall charge but carries both a negatively charged group (carboxyl, $CO_2^-$) and a positively charged group (amino, $NH_3^+$). In proteins the individual amino acids are polymerized, giving a peptide backbone which is electrically neutral except for the ends of the chain. Most of the standard amino acids found in proteins have uncharged side chains, although histidine, lysine and arginine each have a positive charge at neutral pH and both glutamic and aspartic acids normally carry a negative charge.

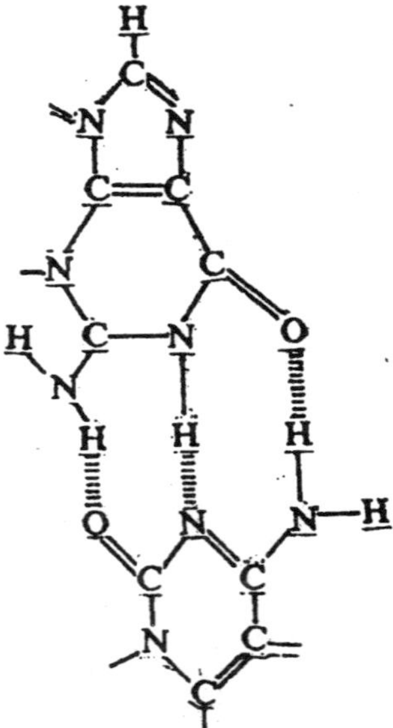

*Fig. 4.7. Hydrogen bonds.*

### *Hydrogen bond*

A second important noncovalent interaction is the hydrogen bond, a dipole formed when a hydrogen atom covalently bonded to an electronegative atom is shared with a second electronegative atom

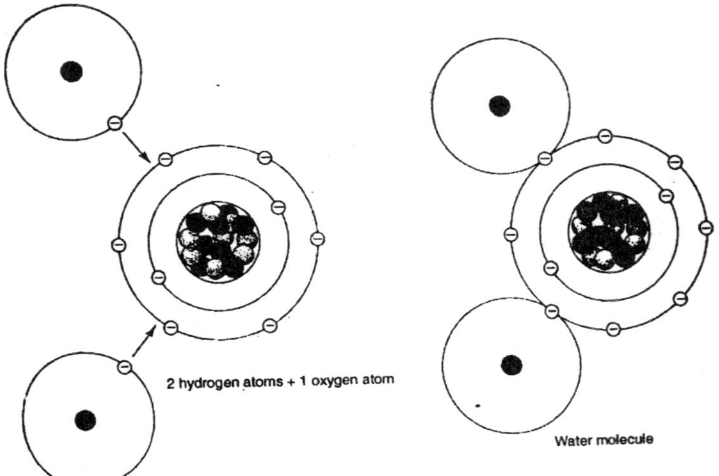


*Fig. 4.8. Covalent bonds.*

(typically an oxygen, nitrogen or fluorine atom), such that the proton may be approached very closely by an unshared pair of electrons. Hydrogen bonds are largely responsible for the unusual thermodynamic properties of water and ice, and the DNA double-helical and protein α-helical and β-structure conformations are extensively hydrogen bonded. Highest bonding energies occur when donor and acceptor atoms are 0.26-0.31 nm apart. Typical hydrogen bond strengths in proteins are 7-50 zJ.

***Van der Waals Interaction***

A third important noncovalent force is van der Waals interactions (London dispersion forces). There is an attractive component due to the induction of complementary partial charges or dipoles in the electron density of adjacent atoms when the electron orbitals of two atoms approach to a close distance. There is also a strongly repulsive component at shorter distances, when the electron orbitals of the adjacent atoms begin to overlap, commonly called steric hindrance. The van der Waals attractive bonding energy between two parallel plates of area A and separation $z_{sep}$ is approximated by:

$$E_{vdW} = \frac{HA}{12\,\pi\, z_{sep}^2} \qquad \text{...(21)}$$

where the Hamaker constant H = 37 zJ for water, 66 zJ for glycerol, 340 zJ for diamond. For A = 0.4 $nm^2$ (small molecule), $z_{sep}$ = 0.3 nm, then $E_{vdW}$ = 4 zJ (water) to 8 zJ (small organic molecules)—close to the mean energy of a thermally excited harmonic oscillator, kT ~ 4.3 zJ at 310 K. While van der Waals bonds are individually very weak, they are also very numerous since they involve all pairs of neighbouring atoms. For example, experimental analysis of an antigen molecule trapped in the anti-hen egg-white lysozyme monoclonal antibody Fv active binding site found 86 distinct interatomic contact points with antigen-antibody separations ranging from 0.25-0.46 nm, averaging 0.36 nm. Since intermolecular dispersion forces act on all molecules, there are probably no ligands with MW > 400 daltons which cannot be receptored. That is, van der Waals interactions ensure that virtually all molecules of nanomedical interest are theoretically bindable noncovalently.

***Aromatic π bonds***

A fourth type of interaction (π electron to π electron), called "aromatic" or "pi" bonding, occurs when two aromatic rings (conjugated π systems) approach each other with the plane of their aromatic rings overlapping, with successive π-bonded systems stacked like layers in a cake. This results in a noncovalent attractive force with a bond strength of ~40-50 zJ. (That is, the planes of conjugated π systems attract each other when superimposed.) π bond stacking forces contribute to nucleic acid stability at least as much as the hydrogen bonds between bases.

***Hydrophobic forces***

Finally, there are the strong hydrophobic forces, due entirely to solvent entropy changes. When two nonpolar residues approach each other, the surface area exposed to solvent is reduced, increasing the entropy of all the water present and decreasing the entropy of the residues, adding to the binding energy a hydrophobic free energy of ~17 $zJ/nm^2$ of contact surface area that was formerly exposed to water.

In designing an artificial binding site, the above forces may be combined to achieve the desired level of affinity and specificity for a given ligand. All forces are not equally useful in this regard, however. For example, hydrophobicity is the major factor in stabilizing protein-protein associations. But hydrophobicity is almost entirely nonspecific, hence contributes little to ligand discrimination. By contrast, the proper formation of hydrogen bonds and van der Waals contacts require complementarity of the surfaces involved. Such surfaces must be able to pack closely together, creating many contact points, and charged atoms must be properly positioned to make electrostatic bonds. Thus van der Waals and polar interactions may contribute little to the dynamic stability of the ligand-receptor complex, but they do determine which molecular structures may recognize each other. Other design elements of binding sites, such as directed channeling of substrates into the receptor, may also prove useful.

In analyzing molecular forces, note that at the nanoscale level, surface/surface, molecule/surface, and molecule/molecule interactions may feature very complicated behaviours. Nanodevices performing work may generate both thermodynamic and mechanical local nonequilibrium conditions, so calculations based on the general forms of interactions and on macroscopic expressions valid at equilibrium conditions should be taken only as basic estimates.

**Ligand-Receptor Affinity**

For a ligand binding to a receptor in a solvent, there will be a characteristic frequency with which existing ligand-receptor complexes dissociate as a result of thermal excitation, and a characteristic frequency with which empty receptors bind ligands as a result of Brownian encounters, forming new complexes, with the frequency of binding proportional to the concentration of the ligand in solution, $c_{ligand}$ (molecules/nm$^3$). For simple processes, the equilibrium constant $K_d$, taken in the direction of dissociation, is:

$$K_d = k_d \,/\, k_a \text{ (molecules / nm}^3\text{)} \qquad \text{...(22)}$$

where $k_d$ is the dissociation rate constant (sec$^{-1}$) and $k_a$ is the association rate constant (nm$^3$/molecule-sec). The $k_a$ rate constant reflects mainly the molecular weight of the ligand, and thus varies little among antibody, enzyme, or other receptor systems. For example, Delaage notes that changing the solution pH for growth hormone from 7.5 to 4.0 increases $K_d$ by a factor of 3000, due to $k_d$ increasing by a factor of 1600 but $k_a$ decreasing only by a factor of 1.7. Hence it is the rate constant of dissociation, $k_d$, which accounts for the vast bulk of affinity in receptor systems.

Thus receptor affinity is usually taken as the inverse of the dissociation rate constant, which may be placed in the context of the half-life of the ligand-receptor complex approximated by:

$$t_{1/2} \sim \ln(2) \,/\, kd \text{ (sec)} \qquad \text{...(23)}$$

Observed half-lives range from <0.1 microsec ($k_d \sim 10^7$ sec$^{-1}$) for the enzyme catalase to a few months ($k_d \sim 10^{-7}$ sec$^{-1}$) for enzyme inhibitors such as the Kunitz inhibitor of trypsin and for avidin-biotin binding. The smaller the $k_d$ (or the $K_d$), the greater the affinity and so the more firmly the receptor grasps the ligand.

The probability $P_{occupied}$ that a receptor will be occupied is given by:

$$P_{occupied} = \left(\frac{c_{ligand}}{K_d}\right) P_{unoccupied} \qquad \text{...(24)}$$

where $P_{unoccupied} = 1 - P_{occupied}$. To ensure $P_{occupied}$ = 99% receptor occupancy, $K_d$ must $\sim c_{ligand}$ / 100. For target molecules present at the $10^{-3}$ - $10^{-11}$ gm/cm$^3$ concentrations typically found in human blood, $c_{ligand} = 3 \times 10^{-3}$ molecules/nm$^3$ for glucose to $c_{ligand} \sim 10^{-11}$ molecules/nm$^3$ for female serum testosterone, giving a range of $K_d \sim 10^{-4}$ - $10^{-13}$ molecules/nm$^3$ to achieve 99% occupancy.

How much binding energy per receptor will this require? The free energy of dissociation $\Delta G_d$ of a ligand-receptor complex is related to its equilibrium dissociation constant $K_d$ by:

$$\Delta G_d = -kT \ln(K_d \,/\, K_0) \qquad \text{...(25)}$$

which refers to a standard reference state where all chemical species are 1 M (i.e., $K_0 \sim 0.6$ molecules/nm$^3$) and attributes a free energy of zero to a complex with a dissociation constant of 1 M.

For T = 310 K, the range of required $K_d$ gives a range for $\Delta G_d$ of 39.4 zJ for glucose (at typical serum concentrations) to 128 zJ for female serum testosterone.

However, when ligand and receptor associate there is a loss of three degrees of freedom in each of translational and rotational entropy, which may be estimated using the classical Sackur-Tetrode equations, giving an entropic free energy range (for translation and rotation combined) of $\Delta G_s$ = 80 zJ for very small molecules (MW ~10 daltons), to 120 zJ (MW ~$10^2$ daltons), 200 zJ (MW ~$10^4$ daltons), and 280 zJ for large molecules (MW ~$10^6$ daltons).

Thus to form a ligand-receptor complex with a dissociation constant $K_d$, the receptor design must provide a free energy of binding of at least:

$$\Delta G_{total} = \Delta G_d + \Delta G_s \qquad \text{...(26)}$$

or 120-410 zJ/molecule for designed receptors achieving 99% occupancy operating over the likely range of physiological concentrations and temperatures. This is consistent with Drexler's estimate of 161 zJ binding energy required to ensure reliable receptor occupancy for small plentiful molecules.

**Ligand-Receptor Specificity**

While affinity measures the strength of the binding of a ligand to a receptor, specificity defines the degree to which a receptor can distinguish between similar ligands. That is, the affinity of the target molecule for the receptor must be greater than the affinity of any other ligand in the environment that is competing for that same receptor, by some threshold multiple.

How much greater is enough? In natural dynamic cellular systems, the threshold multiple appears to be a factor of ~$10^2$-$10^3$. For example, the carrier which expels $Ca^{++}$ from erythrocytes presents a variation in $K_d$ of $10^{-6}$ to $10^{-3}$ in going from the interior to the exterior of the cell. The active transport of amino acids by hepatocytes normally involves $K_d$ ~$10^{-1}$, but under conditions of deprivation a high affinity carrier with $K_d$ ~$10^{-3}$ comes into play.

However, the key to assessing specificity in the nanomedical context would be to tally all the competing molecules in the *in vivo* environment, determine which are the nearest competitors, and then design to avoid them by imposing appropriate energy barriers. By 1998, only a very few competitive ligand-receptor binding analyses had been performed. Until a complete molecular inventory of the human body becomes available, the following crude estimate of nearest-neighbour differences must suffice.

The human body contains a minimum of $N_{prot} \sim 10^5$ distinguishable proteins. The maximum number of distinguishable proteins in the biosphere was given by Kauffman who estimates a useful biological catalytic task space of $N_{prot} \sim 10^8$ distinct protein forms, a tiny subset of the $\sim 20^{500}$ possible 500-residue protein sequences. If the average protein is constructed from $N_{residue} \sim 500$ amino acids (MW $\sim$50,000 daltons), then the average protein may differ from its most similar neighbour by $n_{var} \sim \log(N_{prot})/\log(N_{residue}) = 1.9$-$3.0$ residues. (The precise magnitudes of $N_{prot}$ and $N_{residue}$ are not crucial to our conclusion.) Most proteins are confined to cells containing $N_{prot} \sim 5000$ different protein types each; given that evolution has probably optimized local specificity to ensure that closely competing crucial ligands rarely appear in the same cell, it seems reasonable to assume that the average closest neighbour ligand may differ from the average target ligand by at least $n_{var} \sim 1$ residue.

How much is receptor affinity reduced when binding molecules differing by $n_{var} \sim 1$ residue from the target ligand on receptor-accessible surfaces—the minimum threshold required to ensure specificity within a cell? In one experiment the relative affinity of an antibody constructed for succinylglycinamide-linked histamine (histamine-SGA), which was the target molecule, and the same molecule but with one methyl group or one carboxylic group removed and replaced with a hydrogen, was $1.45 \times 10^4$ or $2.5 \times 10^5$, respectively, due to steric hindrance. Similar investigations with antibodies for SGA-linked serotonin produced relative affinities of 500-1000, and with antibodies for single alanine substitutions in Human Growth Hormone (HGH), $\sim$1000. The relative affinity of a particular RNA oligomer for theophylline and caffeine, two ligands which differ by only a single methyl group, was measured experimentally as 10,900. Computational receptor experiments in which CH replaces N suggest a decline in relative affinity of $\sim 5 \times 10^4$. Indeed, the change of a single hydrogen atom on a ligand is usually sufficient to destroy its specificity for, or activity within, a particular enzyme. The single-residue affinity reduction at a receptor-accessible surface appears to be of order $\sim 10^3$ - $10^5$.

Each increase of $\sim$10 zJ in bonding energy causes the reaction equilibrium constant to decline (hence receptor affinity to rise) by a factor of $\sim$10. If a difference in affinities of $\sim 10^3$-$10^5$ between the target molecule and its nearest competitor likely to be present in the environment provides sufficient receptor specificity for nanomedical purposes, this requirement corresponds to a binding energy differential affinity of ~30-50 zJ at 310 K between target and closest neighbour ligands.

**Ligand-Receptor Dynamics**

Diamondoid structures can exhibit a stiffness and rigidity one or two orders of magnitude greater than that available in protein structures. In general, stiffer structures permit greater specificity because they enhance exclusion of non-target ligands based on van der Waals overlap forces (called steric hindrance) and allow narrower tolerances in distinguishing acceptable ligands.

In less-stiff protein-based receptors, each of the atoms is engaged in relatively large, rapid jiggling movements. Experimental and theoretical work has been done on the atomic fluctuations within the basic pancreatic trypsin molecule, a small enzyme with 58 amino acids and 454 heavy (non-hydrogen) atoms. This work established that fluctuations increase with distance from the center of the molecule, with the magnitude of RMS fluctuations ranging from ~0.04 nm for backbone atoms to ~0.15 nm for the ends of long side chains (roughly one atomic diameter), and an average of 0.069-0.076 nm per atom over the entire molecule. A similar experimental analysis of reduced cytochrome c, a common metabolic enzyme, shows that RMS fluctuations of each of the 103 amino acid residues in the molecule averages ~0.11 nm with lattice disorder (~0.05 nm) included, and fluctuations range from 0.09-0.16 nm. Antibody core domain movements display RMS fluctuations of 0.04-0.19 nm. Hence it appears that the average atom within the typical protein receptor oscillates ~0.1 nm every ~$10^{-12}$ sec, although frequently residues with long side chains (e.g., arg, lys) have much higher RMS deviations than average.

By contrast, in stiff diamondoid-based receptors each of the atoms is locked in a rigid crystalline structure and thus is subject to thermal displacements approximately 10 times smaller. The RMS displacement for a quantum mechanical harmonic oscillator is given by:

$$\Delta X = \left[\left(\frac{\hbar\omega}{k_1}\right)\left(\frac{1}{2}+\frac{1}{e^{\hbar m/kT}-1}\right)\right]^{1/2} \qquad \ldots(27)$$

where $\hbar$ = 1.055 x $10^{-34}$ joule-sec, kT = 4.28 zJ at T = 310 K, and angular frequency w = $(k_s/m_{red})^{1/2}$ rad/sec where $k_s$ is mechanical stiffness and $m_{red}$ is the reduced mass = $m_1m_2/(m_1 + m_2)$. For C-C atoms (e.g., in the receptor body), $k_s$ = 440 N/m, $m_1$ = $m_2$ = 2 × $10^{-26}$ kg, thus w = 2.1 × $10^{14}$ rad/sec, so the RMS displacement of each atom is only ~0.005 nm every ~3 × $10^{-14}$ sec. For C-H atoms (e.g., on the hydrogen passivated receptor surface), $k_s$ = 460 N/m, $m_1$ = 2 × $10^{-26}$ kg (C), and $m_2$ = 1.673 x $10^{-27}$ kg (H), thus w = 5.5 × $10^{14}$ rad/sec, so the RMS displacement of each atom is ~0.008 nm every ~1 × $10^{-14}$ sec. Similarly, at 310 K the RMS thermal displacement of a 1-nm wide, 10-nm long diamondoid rod is ~0.01 nm, including elastic and entropic contributions.

The ratio of RMS displacements for protein/diamondoid receptors is ~10:1, so the minimum addressable volume (hence inverse maximum specificity) of a diamondoid receptor should be ~$10^3$ smaller than for protein receptors, a ~30 zJ binding energy advantage for diamondoid receptors.

**Diamondoid Receptor Design**

Natural enzymes and antibodies are proteins folded into highly organized, preformed shapes that present a ready-made "keyhole" into which a target ligand will fit. The enzyme is folded in such a way as to create a region that has the correct molecular dimensions, the appropriate topology, and the optimal alignment of counterionic groups and hydrophobic regions to bind a specific target molecule. Tolerances in the active sites can be narrow enough to exclude one isomer of a diastereomeric pair. For example, D-amino acid oxidase will bind only D-amino acids, not L-amino acids.

These "keyholes" are extremely floppy, yet still achieve fair specificity. This is a consequence of "induced fit" in protein binding sites. That is, the interaction of the target molecule with an enzyme induces a conformational change in the enzyme, resulting in the formation of a strongly binding site and the repositioning of the appropriate amino acids to form the active site. The receptor flexes, balloons, hinges, or contracts by 0.05-1.0 nm in just the right places to maximize specificity as the selected ligand enters the site. In some cases such as $O_2$ and CO binding by myoglobin, ligands enter the receptor through a series of temporary voids that appear and disappear in the receptor as ~10 picosec dynamic structural fluctuations. Induced fit can reduce receptivity to undesired proteins that exploit relative geometry by bonding enough to bring portions of their surfaces into alignment with the same receptor sites that bind desired proteins. Folding transitions appear to be the most prevalent and to possess the most possibilities for adaptability or induced fit.

For smaller molecules, it is likely that recognition processes will be relatively inefficient, time-consuming, and more difficult to engineer if they involve a good deal of rearrangement of the receptor's shape.[382] Thus there is considerable interest among chemists in designing artificial receptors that have their cavities already formed into the shape appropriate for the intended substrate. For instance, rigid-cavity "spherand" receptors are exceptionally efficient at binding metal ions. Bowl-shaped molecules such as cryptaspherands, calixarenes, and carcerands can be lined with chemical groups along their walls and with charged groups along their rims to achieve high binding specificity. Self-assembling capsules or "container molecules" made of hydrogen-bonded subunits, capable of limited molecular recognition, have been synthesized. A designed receptor for creatinine was demonstrated in 1995, and in 1998, K. Suslick and colleagues designed metalloporphyrin-dendrimeric artificial receptors that can bind straight, skinny molecules but block out bent or fat molecules. Container molecules with 0.2-0.4 nm portals control entry to their interiors using "French door" and "sliding door" gates and hinges. Active binding sites for small simple molecules such as NO, CO, and $C_2H_4$ are well-known.

Ultimately, receptors will be designed to nanoscale precision and may be constructed using diamondoid materials. Electrostatic, hydrophobic, and hydrogen-bond forces will add immensely to artificial receptor specificity and are essential for binding small molecules. For example, using a 0.2-nm range in a saline environment, 10-40 charge contacts would be required to bind molecules of various sizes and concentrations using electrostatic force alone, which is ~0.5 charge/nm$^2$ over the entire surface of a 60,000 dalton globular protein (vs. ~1 charge/nm$^2$ for the surface of an isolated zwitterionic amino acid). Or, a 7 nm$^2$ hydrophobic cavity having the exact folded shape of the target ligand generates ~120 zJ binding energy as the molecule stuffs itself into the cavity to exclude its surface from solvent water.

But consider a theoretical receptor that employs van der Waals dispersion forces alone. Atoms comprising the typical protein or CHON target molecule in the human body have an average atomic mass of ~6 amu/atom and an average density of 1500 kg/m$^3$, giving a mean molecular volume of ~$6.7 \times 10^{-30}$ m$^3$ per atom in the target molecule. Assume for simplicity a spherical receptor surface that forms a negative image of the surface of the target molecule. The receptor surface lies ~1.5 × (minimum van der Waals contact distance) ~0.3 nm from the perimeter atoms of the target molecule, and completely encloses the target molecule (thus requiring at least one moving part). (Proteins are actually ellipsoidal with a much larger surface area $A_p = 0.111$ $MW^{2/3}$ $nm^2$, than if they were spherical, so figures are conservative for protein binding; proteins typically have a ~60-85% interior packing density.) Dispersion forces alone can provide the minimum required binding energy of ~120 zJ with $A_p \sim 6$ $nm^2$ of contact surface (MW > 400 daltons) at 0.3 nm mean range, and dispersion-force receptors can offer exceptionally high affinities for molecules >1000 atoms.

What about specificity? Given the ability to design diamondoid binding sites to at least localized <0.01 nm tolerances, chirality is readily detected and (purely as a design exercise) it may even be possible to distinguish diatomic nitrogen and oxygen on the basis of size alone. The molecular lengths (major axis) of $N_2$ and $O_2$ are 0.250 nm and 0.253 nm, but the molecular widths (minor axis) are 0.140 nm and 0.132 nm, respectively. (Diatomic molecules get longer and narrower at higher molecular weight.) Thus $N_2$ is distinguishable from $O_2$ on the basis of width (~0.01 nm). With a van der Waals energy well depth of 1.1-1.4 zJ for $N_2$ and $O_2$, in a tight receptor these gases are bound at an energy density corresponding to 3000-4000 atm of pressure.

**Minimum Feature Size and Positioning Accuracy**

Maximum displacement measurement accuracy in nanoscale devices is ~0.01 nm, and RMS thermal displacement in diamondoid bonds is ~0.01 nm at 310 K. Thermal displacements in 10-nm long diamondoid rods are ~0.01 nm at a 1-nm rod width, ~0.02 nm at 0.5-nm width, and ~0.10 nm at 0.3-nm width. However, it is possible to construct components to even narrower tolerances.

For instance, a single C-O bond inserted into a diamondoid rod in a collinear carbon chain extends rod length by 0.1402 nm, the C-O bond length. An adjacent rod into which an N-N bond is similarly inserted is extended by 0.1381 nm, the N-N bond length. By bonding these rods (aligned at one end) it is possible to build diamondoid structures having 0.002-nm features (at the other end), which at 310 K will nonetheless suffer thermal displacements of ~0.01 nm or more. Similarly tiny displacements can be induced in binding cavity surfaces by inserting a foreign atom deep inside the bulk diamondoid structure, causing dislocation strains that decline in magnitude at greater distances from the compositional disturbance.

It is also possible to translate diamondoid components through a picometer step size, much smaller than the unavoidable RMS thermal displacements, using any of several methods; for example:

*Levers*

Consider a 10-nm lever joined to a fixed bar by a pivot at one end, and driven axially by a ratchet interposed between lever and bar at the other end. Ratchet movements translate to smaller displacements at positions along the lever distant from the ratchet. Thus a follower rod attached to the lever 1 nm from the pivot and driven by a ratchet with 0.01-nm steps moves ~0.001 nm per ratchet step.

*Screws*

Consider a 3-nm diameter cylindrical screw with a 1-nm pitch. Rotating the screw through a 0.01-nm circumferential displacement causes the screw to move laterally by ~0.001 nm, which may be transmitted elsewhere in the machine by an attached follower rod. Of course, nanoscale screws or gears are sensitive to the precise cancellation of the potentials and hence cannot be perfectly smooth and circular, producing some unavoidable "knobbiness" under load.

*Gear trains*

Consider a 32-nm diameter worm gear with 1-nm teeth. One rotation of the gear requires 100 rotations of the worm; hence a 0.1 nm displacement applied to the worm produces a 0.001 nm displacement in the gear. More efficient (and coaxial) compound planetary gear trains commonly employed in transmissions achieve displacement ratios up to 10,000:1, a hundred times better than the above example.

*Hydraulics*

Consider a sealed, fluid-filled, tapered pipe. A piston 1000 nm$^2$ in area is mounted at one end; another piston 10 nm$^2$ in area lies at the other end. A displacement of 0.1 nm applied to the smaller piston produces a 0.001 nm displacement in the larger piston. (Here again the finite size of molecules may produce "knobby" performance as fluid particles slip from one stable configuration to the next.)

*Compression*

Consider a rod upon which a compressive force of 100 nN/nm$^2$ (near the maximum diamondoid strength) has been imposed. Affixed to the rod are two crossbars spaced 4 nm apart. If the force on the rod is increased to 101 nN/nm$^2$, the gap between the crossbars compresses by ~0.001 nm.

**Receptor Configurations**

Many different useful receptor configurations may be readily envisioned, of which the following brief descriptions are but a small sample. Note that most large target molecules of nanomedical interest are proteins which are probably floppy enough to permit entry into reasonably open multiply-concave rigid receptor structures; if not, hinges are easily added to the receptor design.

***Imprint model***

Molecular imprinting is an existing technique in which a cocktail of functionalized monomers interacts reversibly with a target molecule using only noncovalent forces. The complex is then crosslinked and polymerized in a casting procedure, leaving behind a polymer with recognition sites complementary to the target molecule in both shape and functionality. Each such site constitutes an induced molecular "memory," capable of selectively binding the target species. In one experiment involving an amino acid derivative target, one artificial binding site per $(3.8\ nm)^3$ polymer block was created, only slightly larger than the $(2.7\ nm)^3$ sorting rotor receptors described by Drexler. Chiral separations, enzymatic transition state activity, and high receptor affinities up to $K_d \sim 10^{-7}$ have been demonstrated, with specificity against closely competing ligands up to $\Delta K_d \sim 10^{-2}$ (~20 zJ).

Several difficulties with this approach from a diamondoid engineering perspective include:

1. A sample of the target molecule is required to make each mold.
2. It is currently unknown how to prepare diamondoid castings.
3. Once the imprint has been taken, the site cannot easily be further modified.

***Solid mosaic model***

In the solid mosaic receptor model, the precise shape and charge distribution of the target molecule is already known. Working from this information, a set of diamondoid components could be fabricated which, when fitted together like a Chinese puzzle box, create a solid object having a cavity in the precise shape of the optimum negative image of the target molecule. The mosaic may contain point charges, voids, stressed surfaces, or dislocations to achieve fine positional control. Mosaic components may be as small as individual atoms, so this model is conceptually similar to 3-D printing or raster-scan techniques in which the desired cavity formation is constructed atom by atom inside a nanofactory. This model, like the imprint model, cannot easily be reconfigured once it has been constructed because each of the many unique parts may contribute to the entire structure. The construction of receptors from parts of fixed size and shape is crudely analogous to members of the heterodimer receptor class (a two-component receptor) such as $GABA_B$.

M. Reza Ghadiri has designed a protein mosaic model using cyclic peptides that assemble spontaneously into nanotubes of predefined diameter; incorporation of hydrophobic amino acid side chains on the outside of these tubes leads to spontaneous insertion into bilayers, allowing the tubes to function as transmembrane ion channels. Other examples of mosaic model receptors are mesoporous silica filters with functionalized organic monolayers forming 36 nm sievelike pores, and zeolites and zeolitelike molecular sieves. Zeolites are artificial crystal structures with precise and uniform 0.4-1.5 nm internal void arrays which can also be used as shape-selective catalysts able to favour one product over another that differs in size by as little as 0.03 nm, such as pxylene and oxylene. By 1998, rational de novo computational design of artificial zeolite templates and crystal engineering had begun.

***Tomographic model***

In the tomographic receptor model, the receptor engineer again starts with a known target molecule topography and designs a series of thin planar sections which, when stacked together in the correct order (using positionally-coded docking pins) and bonded, create a solid object containing the desired optimum binding cavity. As in the mosaic model, point charges or dislocations in each planar segment can be used to manipulate cavity features and dimensions to precise tolerances. Unlike the mosaic model, a tomographic receptor can be reconfigured by partial disassembly and replacement of specific planar segments, each of which contributes only locally to the total receptor structure. Hybrid or modular artificial enzymes and two-dimensional sheetlike hydrogen-bonded networks are crude analogs in current research.

***Pin cushion model***

The pin cushion receptor, suggested independently by K.E. Drexler, is a hemispheroidal or hemiellipsoidal shell through which a number of rods protrude, each of which may be moved radially. When inserted through the shell to varying depths, the endpoints of the rods define a negative image surface which may be made to mirror the topography and charge distribution of a known *target molecule*. Rods may be tipped with positive, negative or no charge, or they may terminate in any number of functionalized surface segments designed to optimally match parts of the target molecule shape. Other configurations such as a rectangular box, hinged plates with protruding rods, counterrotating rollers, or time-varying rod positioners are readily conceivable. Pin cushion receptors are easily reconfigured to bind different target molecules, hence may be regarded as fully programmable "universal" binding sites. The principal difficulty with the pin cushion receptor is its excessive size (compared to other receptor models) and its greater complexity (since each rod may be controllable individually).

Pin cushion receptors can also be used to discover the shapes of unknown molecules: A target molecule is placed in the central cavity with all rods fully retracted, and the rods are slowly slid forward using nanopistons with force reflection feedback, until all pistons register zero force, indicating balance between attractive and repulsive van der Waals interactions, at which point all rod positions are recorded. Rods of differing end tip charge may then be tested for additional attractive potential. The final result is a precise mapping of the target molecule, which data may be stored or transmitted elsewhere for future use.

***Construction costs***

The active binding site of a receptor consisting of $(2.7\ nm)^3 \sim 19\ nm^3$ of structural atoms and constructed with 0.001-nm feature sizes in theory requires information from $2 \times 10^{10}$ voxels (volume pixels) for complete description. However, there are only $N_{atom} \sim 1000$ atoms involved in the structure and their locations cannot be arbitrarily chosen. Atomic scale ($\sim$0.1 nm) resolution would require $\sim$19,000 voxels; each voxel minimally requires an index number ($\sim \log_2(19{,}000) \sim 14$ bits), an atomic identifier ($\sim \log_2(92) \sim 7$ bits), and a charge identifier ($\sim \log_2(3) \sim 2$ bits), for a total of 23 bits/voxel which gives $4 \times 10^5$ bits/receptor at atomic scale resolution. Drexler estimates the number of configurational options per atom $N_{opt} \sim 150$; hence a description of the receptor could require as few as $N_{atom} \log_2(N_{opt}) \sim 7 \times 10^3$ bits. Assuming energy dissipation of $\sim$3 zJ/bit for rod logic register reading or $\sim$kT ln(2), then each time the receptor description is retrieved, stored or processed may require a minimum energy dissipation of $\sim 10^4$-$10^6$ zJ. Reversible computing may reduce this energy requirement by a factor of 10-100 or more. Construction of a receptor containing $\sim$1000 atoms using the nanomanipulator arm requires $\sim 10^{-2}$ sec and consumes $\sim$0.001 picojoule ($\sim 10^6$ zJ) of mechanical energy.

***Receptor durability***

Receptor durability is difficult to estimate ab initio. Molecularly imprinted polymer sites can be stored at least several years without loss of performance, but these receptors have only been tested to $\geq$100 cycles of use without any detectable loss of memory. The lifetime *in vivo* for metabolic enzymes often exceeds ~$10^5$ sec (~1 day), and taste cell receptors survive ~$10^6$ sec (~2 weeks), which suggests operational lifetimes for natural receptors on the order of $10^7$-$10^{12}$ cycles despite the relative fragility of protein structures. Diamondoid structures should be even more durable because of their superior physical strength, affirmative forcible ejection of bound ligands each cycle, and high resistance to chemical degradation thus reducing susceptibility to poisoning.

**Ligand-Receptor Mapping**

To achieve a fully general-purpose receptor capability in nanomedical systems, two classes of analytic function are essential.

First, presented with an arbitrary molecule, the system or user must be able to infer from the molecule's structure the shape and electronic configuration of an optimal receptor geometry that will efficiently bind it, with a particular affinity and specificity (as a design specification). This discovery procedure may involve a process akin to molecular imprinting, fluorescent dye affinity matching on testing chips, Structure Activity Relationships (SAR) by NMR techniques, or pin cushion receptor mapping.

Second, presented with an arbitrary protein-built binding site embedded in living tissue, the system or user must be able to infer the molecule(s) which the given receptor could bind, and to compute the affinity and specificity of that activity. This capability may require a rather diverse steric toolkit. Biological receptors are constructed in one of ~500-1000 distinct shapes or "domains," such as the well-known Y-shaped antibody immunoglobulin domain (~100 residues) and other assorted clefts, folds, kringles and coils. Several hundred distinct fold families are known, though it is believed there are only ~20 major domain types, and Chothia believes that "the large majority of proteins come from no more than one thousand families."

One mapping technique would employ a series of rodlike probes inserted into the receptor cavity like a pick gun or other lockpicking tool. After physically securing the receptor, the first crude probes quickly map the cavity to nanometer scale. Based on this preliminary information, subsequent probes having ever-finer discrimination chart smaller features as well as charge distributions by inserting a predetermined set of test rods with functionalized tips in a standard sequence, to rapidly prune the huge configurational space down to a single unique electrophysical shape using the minimum possible number of tests. (A 1 $nm^3$ volume of multi-element diamondoid receptor structure has $\sim 10^{148}$ possible distinct configurations, by one conservative estimate, requiring at least 492 binary tests to eliminate all but one configuration. Antibody domains contain $\sim 10^{50}$ possible configurations, requiring 166 binary tests.) Cavity Stuffer, a software package comprising an experimental design tool to investigate automated cram-packing of predefined cavities using randomly branched polymers, is a preliminary effort in this general direction, although the algorithmic task of discovering an unknown receptor contour may be considerably more challenging.

Another approach to receptor mapping would be the reversible chemical or mechanical denaturation of the receptor protein followed by precise nondestructive amino acid sequencing, from which tertiary structure and activity could then be computationally inferred. Algorithms to perform such computations are the subject of intense current research interest. Even imperfect tertiary structure predictions should greatly reduce the search space, so that only a partial residue sequencing may be necessary for unambiguous identification from a library of possible proteins. Once the target ligand structure has been inferred, the subsequent design and manufacture of receptor-specific agonists and antagonists, including catalysts and cofactors, activators and inhibitors, promoters and repressors, should be comparatively easy.

**Large Molecule Binding, Sorting, and Transport**

Is there any size limit for target molecules to be transported? Natural receptors have already been found for large molecules including low-density lipoproteins (LDLs) > 1,000,000 daltons and high-density lipoproteins (HDLs).

The methods described in earlier Sections can be adapted for binding large molecules (>1000 atoms), including molecules far wider than the binding device itself (e.g., ~200-nm diameter virus particles and larger). Making a binding site for a large molecule should be physically easier (albeit computationally more challenging) than making a binding site for a small molecule because of the greatly increased area of interaction. For example, a binding energy of 400 zJ may be realized by creating a dispersionforce binding area covering only ~25% of the surface of a 10,000-atom target molecule or a mere ~0.02% of a 200-nm virus particle.

This makes possible the concept of binding pads—small surfaces with dimples (concave or convex), each dimple consisting of precisely-placed nanometer-scale features that are complementary to specific patches (e.g., epitopes) on the surface of the large target molecule.

(Specificity is lost for portions of the molecule outside the particular patches.) Each dimple could effectively grasp the side of the large molecule without having to fully enclose it—a capability useful in nanorobot foot pads during cell walking and anchoring, in handles for nanociliary or nanomanipulator transport functions, and for chemotactic sensing. (For example, macrophage receptors for LDLs employ a "pad" consisting of three globular cysteine-rich domains.) The large compliance of target protein molecule subunits should prove curative for any misalignment problems caused by cumulative small errors in bond lengths across large diamondoid receptor structures.

Rather than using sorting rotors, which become unwieldy when large molecule binding pockets must be used, the shuttle pump may provide reasonably efficient large molecule sortation and transport. The shuttle pump consists of a diamondoid tube within which a receptor ring moves between iris diaphragms at either end. The receptor ring is constructed as two or more binding pad segments. For molecule pickup, the ring is pressed together, forming an annular binding region for the target large molecule, which binds and is shuttled to the other side. The receptor ring is then fragmented, destroying binding affinity and unlocking the target molecule, which escapes via diffusion. The shuttle returns to the pickup side, the receptor is pressed together again, and the cycle repeats. A biocompatible solvent environment is maintained during large-protein manipulation tasks.

Assuming a roughly spherical large molecule and laminar fluid flow at 1 atm forcing pressure, a 10-nm diameter molecule moves through a 20-nm long pump ($\sim 10^{-20}$ kg, $\sim 10^{6}$ atoms) in $\sim 10^{-6}$ sec at $\sim 0.02$ m/sec, consuming $\sim 0.02$ pW during transfer. A 200-nm virus-size target molecule moves through a 400-nm long pump ($\sim 10^{-17}$ kg, $\sim 10^{9}$ atoms) in $\sim 10^{-2}$ sec at $\sim 60$ microns/sec, consuming $\sim 10^{-16}$ watts during transfer; at $\sim 0.0002$ atm, release time is diffusion limited. The transfer force exerted on a 10-nm molecule is $\sim 1$ pN, $\sim 600$ pN on a 200-nm virion; a binding energy of 400 zJ at a 0.2-nm contact distance gives a binding force of $\sim 2300$ pN, sufficient to hold a particle of either size firmly during transport and release.

J. Soreff points out that as protein size increases, so does the energy available for local minima in the binding. Desired proteins may become stuck in incorrect positions, or undesired proteins may become partially adhered to a receptor. Besides designing to minimize these possibilities, using a multireceptor cascade with different combinations of binding patches at each stage should allow complete exclusion of undesired large-molecule species.

# 5

# NANOCHIP

Gordon Moore once said that the number of transistors on a microprocessor will double every 18 months. He has been proved right and now semiconductors are becoming smaller and faster, with currently the target being around trillions of transistors on a microprocessor. This phenomenon did not restrict itself to the semiconductors. It has induced similar developments in other elements that help in proper functioning of the semiconductors. And one of the fundamental essential elements is the power supply.

Electronic devices have shrunk in size substantially over the years and now they are entering the wireless era. The gadgets are getting smaller and at the same time performing many more functions. With mobility becoming a major feature of the present generation devices, the power supplies are expected to match the requirements of these devices. This has essentially led to large investments of time and money in research that would develop better portable battery technologies for mobile applications and also for nanotechnology applications.

The change in the portable power technology has been more evolutionary than revolutionary. Researchers are looking at gaining more out of the basic principal of extracting electricity from chemical energy. The role of material technology is noteworthy in this regard, better materials have given increased results in terms of power density and stability. Researchers are concentrating more on finding the best chemical combination that would provide a much stable and high density power. Historically, batteries have always contributed a major portion of the overall device weight. The latest attempts have been directed towards reversing this without affecting the battery's capability or efficiency.

Researchers have come a long way in reducing the size of the power supplies, and now they are looking at integrating it with the devices. Few of the developments in this direction are illustrated in this article.

## Power Plants on a Chip

With an objective of integrating the portable/miniature power source into the hand-held electronic device, a few universities have invented ways of embedding them into a silicon chips. These include the following:

### Lithium-based

One of the major research working in this area has been undertaken by Hosei University in Japan. The principle of power generation is the conventional method of using an electrolyte and electrodes. Trenches with dimensions 200- by 100- by 2-micron are etched into the silicon chips to house many batteries. These trenches on the chip are filled with the mixture of porous glass electrolyte and the electrodes made from lithium and lithium manganese oxide. Nano-sized pores were added to the glass surface to open up more paths for the lithium ions to travel thus increasing the power of the tiny battery. These batteries generate 3.6 volts and deliver a power of 34.6-watts per square centimeter.

Researchers are further trying to increase the output by embedding more number of smaller batteries into the silicon chip. These batteries are expected to go commercial in five to ten years, and would find their application in computer chips and biochips.

### Fuel Cell-based

A set of researchers from Lehigh University, Pennsylvania, USA. have built a silicon chip device that acts as a fuel cell reformer. The reformer works on the same principle as that of the conventional fuel cells, but it is embedded into a silicon chip. The reformer extracts the hydrogen from the fuel mixture and supplies it to the fuel cell reactor to combine with oxygen and release energy.

The Lehigh device has tiny capillaries etched into the silicon chip, and are meant to transport the raw materials (methanol and water) needed for the reaction. These capillaries lead to a channel which is coated with the catalyst layer of copper that would cause the reaction. These channels are 1 to 2 centimeters long and half a millimeter deep, while the copper coating is about 33 nanometers thick. Once the methanol-water combination reacts with the catalyst, they release the hydrogen. This hydrogen is then supplied to the micro reactors of the micro fuel cells to produce electricity.

The fuel mixture of methanol and water was selected as a solution to overcome the lack of stable and high density fuel. Hydrogen stored in the form of gas or metal hydrides does not provide a high density of energy a compared to the hydrocarbon liquids. And the storage of liquid hydrogen needs high pressure vessels, which would add weight to the power source.

This device is said to have an excellent potential for mass production and is practically viable for commercialization in the future. In the future, fuel cells are going to be a major source of renewable energy for the electronics industry. Fuel cells are quite popular in the industry as they are completely environment friendly.

Although a breakthrough innovation, this product comes with its own problems of having a proper control mechanism. With the miniature size, it is difficult to have a regulated inflow of fuel and it also becomes difficult to monitor the performance and the environment of the device.

The methods could eventually be used in chip-sized fuel cells to power devices that require small, rechargeable power sources, like laptop computers and cellular phones.

**Other Breakthrough—Slimmer and Trimmer**

Many other options are being explored for creating batteries that would serve the future applications effectively. One of the area that is being extensively researched is that of thin film technology, which has been making inroads into battery production. Polymers have been extensively used in making this concept a success. Polymers help in mounting the electrodes coating and also packaging all the elements of the battery within the required dimensions. The typical dimension being 5 to 25 micron in thickness.

Many organizations have been successful in constructing thin film batteries using this technology and are planning the commercialization of this product very soon. Few of these organizations include Cymbert Corp and Infinite Power Solution, which plan to place the product in the market very soon, while NASA plans to improve the product for its aerospace application. The batteries typically operate using lithium ion battery principle. The major advantages of these thin film batteries are : no leakage, cheap to produce, versatile (can be used in any application), no toxic content, and good thermal stability. The applications of this battery are wide spread:

1. Implantable medical devices (pacemaker, hearing aid, etc.),
2. Flexible power packs that combine solar energy collection with energy storage all in one lightweight and versatile package
3. Power for microelectronics such as smart identification cards, computer CMOS-RAM, etc. and many other major aerospace applications.

The developments in the electronics/semiconductor field have seen the batteries adapting to their rate of changes. The battery industry is trying to beat the pace of technology improvements in the application field. Power source is a eternal requirement for any device and cannot be done away with, so the industry players are trying to gain over the competition by making their product compatible with the leading applications. With all devices becoming mobile the need for portable supply is enormous and nanotechnology would be a big consumer of miniature portable power supplies. This should drive the manufacturers to produce suitable and efficient power supplies that are portable. At the moment the battle is more on the R&D front rather than the production or sales front, as the mobile and miniature applications are expected to take some time to proliferate in the markets. But when the time arrives the companies must be armed with the relevant technology to beat the competition. The major issue is the convenience of the customer, due to which devices are becoming mobile. This means battery manufacturers also have the obligation to satisfy this need of the customer.

## Fuel Cell Basics

Take a simple physics experiment: send an electric current through water to split the water into hydrogen and oxygen components. Now, run that experiment backwards. Combine hydrogen and oxygen—the result? Water and electricity. Such was the reasoning of Sir William Grove, a British attorney and physicist in the mid-1800s. In 1839, he put this reasoning into the design of his gas voltaic battery, which is acknowledged today as the first fuel cell. And even though other fuels have been used—in 1889, for instance, chemists Ludwig Mond and Charles Langer attempted to build the first practical fuel cell using air and industrial coal gas—hydrogen-based fuel cells remain the primary focus of efforts to make small, practical fuel cells.

1. Fuel cells are electrochemical devices that convert the energy released by a chemical reaction directly into electrical energy. In

a typical fuel cell, a gaseous fuel, such as hydrogen, is fed to the anode. At the same time, an oxidant, such as oxygen from the air, is fed continuously to the cathode. The fuel and oxygen meet at the electrodes, and the electrochemical reaction occurs. The resulting electrical current is delivered through current collectors.

2. Fuel cells are akin to batteries, but different in some fundamental ways. They have many components in common—anodes, cathodes, and so on. And like batteries, fuel cells can be connected together in series to produce higher voltages. However, a battery stores its energy internally and has a fixed amount of material available for conversion to electricity. Once that material is depleted, the battery is useless or must be recharged. In a fuel cell, the fuel is stored outside the cell itself. Hence, a fuel cell has no fixed capacity—it will generate electricity as long as it is supplied with fuel and air.
3. Fuel cells can accept almost any kind of fuel, including gases such as hydrogen and methane, liquids such as gasoline, and solids such as carbon. Once connected to a fuel supply, a cell will produce electricity until its supply is removed or exhausted.
4. Fuel cells have been used in spacecraft and military applications for years. Livermore researchers are developing a number of types of fuel cells for different applications, including solid-oxide fuel cells, Solid-Oxide Fuel Cells Stack Up to Efficient, Clean Power), carbon conversion fuel cells. Turning Carbon Directly into Electricity), and the fuel cells described in this article, which use a proton-exchange membrane to combine oxygen and hydrogen to create electricity and water—a direct descendant of Grove's first fuel cell.

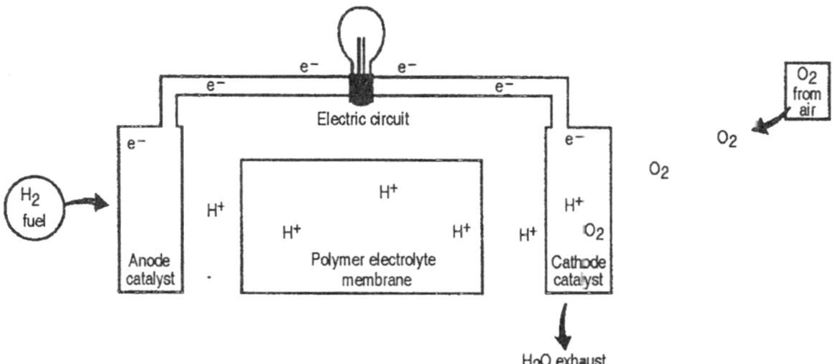


*Fig. 5.1. Anatomy of a micro fuel cell.*

## To Grow a Biological in vitro Power Source on a Chip

There is much interest in finding alternative and renewable energy processes. Significant gains have been made in approaches, such as photovoltaic, wind, and solar/thermal, but the costs of these units can still be considerable; and some require rather sophisticated manufacturing infrastructure. A biologically driven energy source is an appealing alternative, especially one that could convert waste into energy.

### Problem

Consider the design of a power source that is biological in nature and provides an energy output that can be utilized reasonably in an industrial setting (i.e., electricity, hydrogen). This system does not have to be suitable for in vivo use, nor does it have to rival the absolute efficiency of conventional systems; but it should have the potential of improving the current costs to produce clean energy. For comparison, a current, commercially available, solar panel of 1.27 $m^2$ can generate 167 watts with irradiation of 1 $kW/m^2$ (~13 percent efficiency) and costs about $600 to purchase.

As one example of biological power sources, photosynthesis is used by plants to convert water and carbon dioxide into ATP and carbohydrates This cycle can be interrupted to produce hydrogen. Other microorganisms (bacteria, algae) can also be used to produce hydrogen, in a similar cycle. Unfortunately, most of these are self-limiting reactions, where the organisms are inhibited by the reaction by-products. It may be possible to modify the enzymes, or the sensitivity of the organisms to the reaction by-products, to improve the efficiency of these processes, but other technical approaches (i.e., the use of membrane reactors, scavengers, etc.) have also been attempted at the macro scale. Micro and nano scale systems offer many advantages for the design of continuously operating systems for biological energy conversion.

### Solution

A handful of engineers, a pair of biologists, a couple of chemists, and a sociologist walk into a conference room. They've been told to solve a problem. One of the biologists says to the others, effectively, "This problem is stupid." Debate ensues and, in the end, two probably patentable ideas emerge, one posed by the dissenter.

The punch line of the story isn't laughable, but there is a punch line, nonetheless: diverse and intelligent minds stuck in a room together can do a lot in 8 hours. This focus group was tasked with the problem titled, "Grow a biological in vitro power source on a chip," which implied to some group members a small-scale application, perhaps powering an implantable medical device.

But below the title, the problem description explained the goal as a technology that "should have the potential of improving the current costs to produce clean energy." This description implied to other group members that the aim was to devise a large-scale alternative energy source-replacing photovoltaics, for example—based on biology.

The researchers' first job, then, was defining the problem. They brainstormed broadly and discussed both approaches, initially responding to the skepticism offered by one group member that a small biological power cell was a "fundamentally non-doable, non-worthwhile problem."

After initial brainstorming, the group began, somewhat inadvertently, by designing a system that achieved the latter goal of a large-scale biological power source. But by the end of the four 2-hour sessions, the group had design ideas for both applications.

In brainstorming and defining the problem, the researchers discussed power requirements for various devices and applications, and considered the unique attributes of biology and the nanoscale in energy conversion. In particular, group members considered the efficiency of photosynthesis—nature's way of converting renewable, light energy into energy for growth. Photosynthesis is about 35 percent efficient at harvesting light energy (at certain wavelengths), but only about 1 percent efficient at converting that energy into glucose. By comparison, a modern photovoltaic panel is about 15 percent efficient. The group discussed possible applications that could accommodate the mediocre efficiency of photosynthesis.

The group also discussed unusual examples of energy generation in nature, including the electric eel, capable of producing a single 600-volt shock each hour. They discussed microbiological approaches to energy conversion, noting the disadvantage that microorganisms are evolved to use harvested solar or chemical energy for growth, not for surplus power generation. This makes such systems inefficient at external power generation, and can cause devices to foul as organisms multiply and form biofilms.

By the end of the first session, the group had discussed both ideas and sketched the architecture of a plan for a large-scale biological power source. The goal of creating a small, implantable, biological power source initially seemed intractable, because calculations suggested supplying enough power to run anything for a useful period of time was unlikely to be possible. Later, the realization that some devices have very low power requirements opened the door for discussion of this as a feasible approach.

The second session included a more detailed mapping of a solution to the large-scale power generation problem, making use of the expertise of several of the group members. With the initial concept—posed by the group's initial skeptic—on the table, each researcher contributed to the solution. Leaving the initial debate behind, the group grew enthusiastic as ideas came together.

The group's proposal circumvents the low efficiency of natural photosynthesis by converting light directly into electricity, eliminating a carbohydrate intermediate. The approach includes a strategy for expanding the spectrum of absorbable light beyond the normal (narrow) range allowed by photosynthesis.

One of the group members said of the proposed approach, "I think it stands scrutiny as an idea."

Returning to the room the second day, the group discussed the small-scale power generation device. Their approach mimics power generation by the electric eel, which has long fascinated scientists. Looking for information on-line, the group found Volta's 18th century drawings of the eel's electric organ and descriptions of surprising early experiments.

As the sixth hour in the room approached, the group allowed itself to joke about Star-Trek-like possibilities, including implantable eels, noting that if body piercing could take off, so could eel implants.

In the final session, late Saturday afternoon, conference keynote speaker Dr. Mark Humayan, of the University of Southern California, joined the group before his evening talk on the implantation and testing of an artificial retina in patients who had lost their vision. He brought medical expertise to the group, and his knowledge of power requirements for medical implants—which are much lower than the group thought—allowed the group to identify a feasible application for their small-scale power-generating device and to elaborate on its design. Again, enthusiasm grew as the group approached a solution, and the broad expertise of the researchers was impressive to see as they contributed to the solution.

Throughout the discussions, the group remained focused on quantitatively evaluating the feasibility of their ideas: back-of-the-envelope calculations flew fast and furious. The researchers recalled values from memory for biological and physical parameters—power requirements, dimensions, absorption spectra, process efficiencies—and fluently manipulated these numbers to estimate the limits of possible approaches and the requirements of possible applications.

Both environmental and health benefits exist for society by the development of these ideas. A biological power source would offer a clean and renewable energy source, avoiding fossil fuel consumption and the need for toxic materials used in photovoltaics and batteries. A small, implantable biological power source would be biocompatible and alleviate concerns about implantation safety and disposal of today's batteries, which contain metals and other highly toxic components; modern implanted batteries must be carefully encased before implantation.

# 6

# NANOTECHNOLOGY OF DIAMOND

Diamond and diamond-like materials promise to play a large role in advanced molecular nanotechnology. One of the goals of advanced molecular manufacturing will be the use of highly reactive molecules to build up diamond structures and diamond-like structures of the sorts that we have seen in gears, bearings, and many other areas. Because diamond is such an extraordinary material, there is growing interest and commercial activity in the synthesis of diamond today (with existing methods), which shed some light on both the utility of diamond and on how it can be made. Ralph Merkle characterized diamond as an all-around wonder material.

Some of the outstanding and unique properties of diamond are listed in Table 6.1. It is absolutely an extraordinary material in every respect. If you could use any material you wanted to for most engineering tasks, diamond is what you would choose.

For example, in tooling, the hardness, tensile and compressive strength, as well as low thermal expansion of diamond make it an excellent material. More recently, we have learned to appreciate its high thermal conductivity. All of these properties of diamond combine to give cutting tool performance that is at least a factor of 50 better than the best available conventional technology today.

If diamond is such a great material, why is it not more abundant? Diamond is only thermodynamically stable at huge pressures and relatively high temperatures. If you try to crystallize carbon under normal conditions, you wind up with very nice single-crystal graphite. During the early 1950s, General Electric and a Swedish laboratory made parallel discoveries of how to imitate nature and grow diamond

**Table 6.1. Properties of Diamond**

*Diamond physical properties*

| *Property* | *Diamond's value* | *Comment* |
|---|---|---|
| Chemical reactivity | Extremely low | |
| Hardness ($kg/mm^2$) | 9000 | CBN: 4500 SiC: 4000 |
| Thermal conductivity (W/cm-°K) | 20 | Ag: 4.3 Cu: 4.0 BeO: 22 |
| Tensile strength (psi) | $0.5 \times 10^6$ (natural) | $14 \times 10^6$ (theoretical) |
| Compressive strength (psi) | $14 \times 10^6$ (natural) | $80 \times 10^6$ (theoretical) |
| Thermal expansion coeff. ($°K^{-1}$) | $0.8 \times 10^6$ | $SiO_2$ $0.5 \times 10^{-6}$ |
| Refractive Index | 2.41@590 nm | Glass: 1.4-1.8 |
| Transmissivity | 225 nm - far IR | Widest known |
| Coeff. of Friction | 0.05 (dry) | Teflon: 0.05 |
| Band gap (eV) | 5.4 | Si: 1.10 GaAs: 1.43 |
| Electrical Resistivity (Ω-cm) | $1 \times 10^{16}$ (natural) | AlN: $1 \times 10^{14}$ $Al_2O_3$: $1 \times 10^{15}$ |
| Density ($gm/cm^3$) | 3.51 | |

*Diamond semiconductor comparisons*

| *Measurement* | *Si* | *GaAs* | *Natural diamond* |
|---|---|---|---|
| Band gap (eV) | 1.10 | 1.43 | 5.45 |
| Hole mobility ($cm^2/V$-s) | 600 | 400 | 1600 |
| Electron mobility ($cm^2/V$-s) | 1500 | 8500 | 1900 |
| Breakdown field (V/cm) | $5 \times 10^6$ | $6 \times 10^6$ | $\geq 1 \times 10^7$ |
| Electrical resistivity (Ω-cm) | $1 \times 10^3$ | $1 \times 10^8$ | $1 \times 10^{16}$ |
| Work function (eV) | 4.8 | 4.7 | 4.8 |
| Carrier Lifetime (s) | $2.5 \times 10^{-3}$ | $10^{-8}$ | Unknown |
| Thermal conductivity (W/cm·°K) | 1.45 | 0.46 | 20 |
| Electron velocity (cm/s) | $1 \times 10^7$ | $1 \times 10^7$ | $2.7 \times 10^7$ |
| Dielectric constant | 11.0 | 12.5 | 5.5 |
| Lattice constant (Å) | 5.43 | 5.65 | 3.57 |
| Hardness ($kg/mm^2$) | $1 \times 10^3$ | $6 \times 10^2$ | $1 \times 10^4$ |
| Refractive index | 3.5 | 3.4 | 2.41 |
| Thermal expansion coeff. ($°K^{-1}$) | $2.6 \times 10^{-6}$ | $5.9 \times 10^{-6}$ | $0.8 \times 10^{-6}$ |
| Melting point (°C) | 1420 | 1238 | N/A |

by various catalyst and solvent-driven processes in the region. The form in which that process makes diamond available is pretty limited—it is grit. Nonetheless, because of the enormous performance advantage of diamond over other types of abrasives, there is a half-billion-dollar-per-year industry based upon crystal growth at 100 kilo bars and 1,300°C.

## Synthesizing Diamond

The answer to the problems raised by the limitations of the early methods for synthesizing diamond has more recently become available

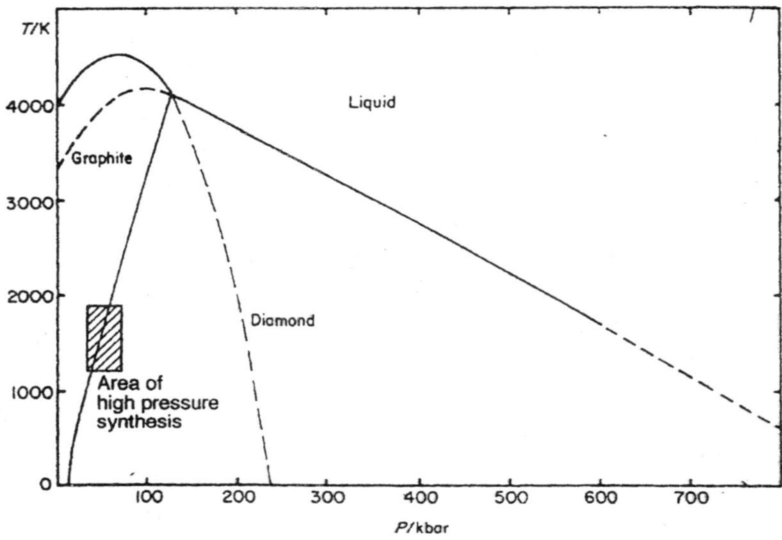


*Fig. 6.1. Phase diagram of forms of carbon.*

to us. At very low pressures compared to the classical diamond synthesis (that is, low vacuum to moderate atmospheric pressures), it is only necessary to take a mixture of hydrogen and some carbon source, and to activate the gases. The carbon source can be nearly anything—benzene, Kirin beer (!), swamp gas, anything with a carbon atom. The gases are activated by placing energy into them in one way or another. This can be done with electrical discharges (which is how we usually do it at Crystalline) with DC glow discharges and microwave plasmas. It can also be done by purely thermal means. In fact, if we heated the tungsten filaments in the light bulbs here to a couple of thousand degrees centigrade, filled the bulbs with hydrogen and methane, we would be diamond-plating the inside of the light bulbs. It is literally that simple. However, do not try this at home. This is for trained professionals only. A hydrogen-oxygen explosion caused by a leaky system is a high-energy event not to be wished for.

There are also other excitation methods that can be used. The basic reason for putting excitation energy into the gas mixture is to create at least the two species. These are the principal suspects for what is going on at the diamond growth surface. If you generate atomic hydrogen and methyl radicals, and put a properly prepared substrate next to the activation zone, you nucleate diamond and grow the diamond. It is that simple. The product is a polycrystalline diamond film. Each

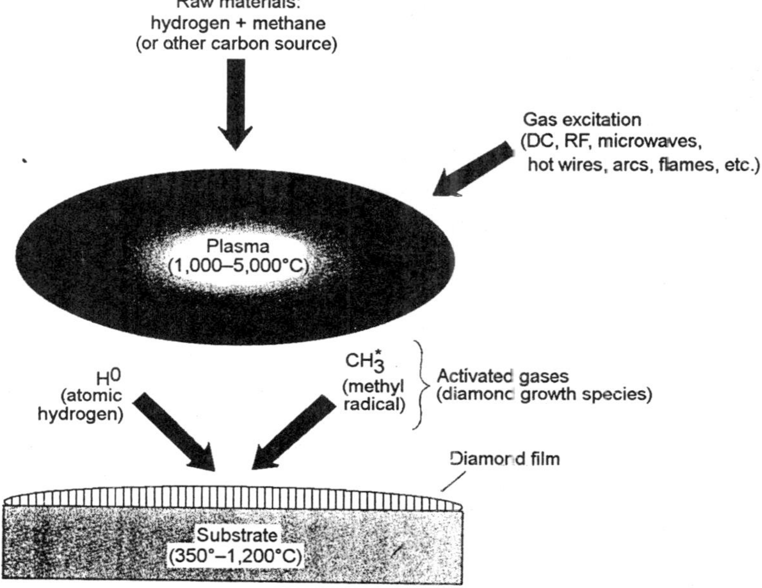


*Fig. 6.2. CVD diamond deposition process with hydrogen and methyl radicals.*

of the domains is a single diamond crystal. In this situation, the individual crystals are not oriented, although predominantly ill texturing is seen. This means that the pyramid-like crystals show primarily triangular facets, which are 111 surfaces. Raman spectroscopy is a

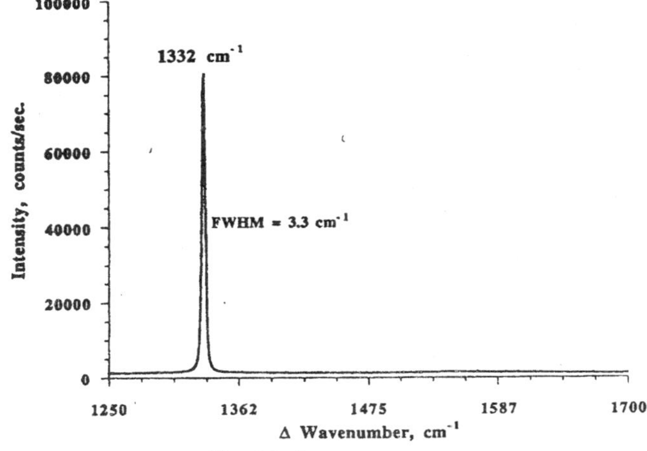


*Fig. 6.3. Raman spectrum.*

nondestructive optical technique that allows an inventory to be made of the chemical bonding state of the carbon atoms. The single sharp peak at 1332 $cm^{-1}$ and the near absence of peaks in the 1550 $cm^{-1}$ region indicates a very high-quality film that is predominantly in the crystalline $sp^3$ (diamond) bonded state of carbon, and very little in either the amorphous or graphite forms of carbon.

Single-crystal natural diamond has a very sharp line, with a width at the point half the height of the peak of only 2 to 3 $cm^{-1}$ which has implications for thermal conductivity, grain size, and various other aspects of the atomic crystal lattice. A diamond-like carbon material that can be made with CVD contains dendritic $sp^3$ or diamond, domains

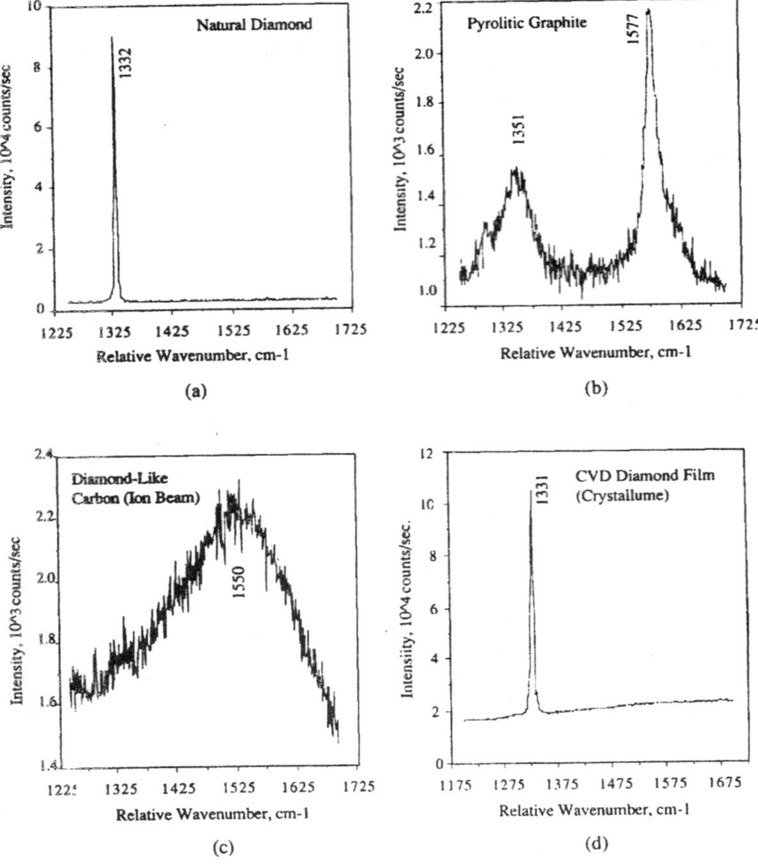


*Fig. 6.4. Raman spectra for different forms of carbon.*

with amorphous or graphitic material filling in between the dendrites. The spectrum shown is typical of microcrystalline graphite, which predominates over the $sp^3$ energy underneath because the Raman scattering cross section for $sp^2$ bonded carbon (graphite) is about an order of magnitude greater than for $sp^3$. Therefore, the Raman technique is differentially sensitive for graphite compared to diamond.

Making a carbon film with ion beams gives a different kind of diamond-like carbon. In this case, the product is mainly amorphous carbon with some $sp^2$ content.

The Raman spectrum is very similar to that of natural diamond. The important point to note here is the structure. It is columnar growth that occurs following nucleation of isolated nuclei, lateral and vertical growth of the nuclei until coalescence occurs, and then vertical growth thereafter.

These structures can be grown on a variety of substrates. The easiest substrate to use is, of course, diamond for *homo-epitaxial growth*. Some of the most difficult substrates include, for example, cobalt-consolidated tungsten carbide tool bits. An easy substrate to use to grow diamond is silicon carbide ceramics or diamond-polished silicon wafers.

The upper panel shows a micrograph of a single 111 face that is about 120 microns on a side. When you see this kind of material in a scanning electron microscope, you know that it has excellent thermal conductivity. The lower panel shows a much smaller-grained diamond film, of the sort that we make for *X-ray lithography* mask membranes. It has rotten thermal conductivity, but it has great structural integrity, and it can be made in 400 nm-thick films over large areas, meaning 4-inch diameter wafers with no pin holes. It is helium leak-tight.

Process parameters for diamond growth that determine the crystallite morphology include carbon concentration, deposition temperature, and current density at the substrate.

## Excitation Mechanisms

One of the technologies that we use at Crystallume is DC Glow Discharge. One advantage of this technology is very high nucleation density, so we can make X-ray spectrometry windows using this technology. Another advantage is that it scales up very easily. To cover larger areas, it is only necessary to buy a larger power supply. A third advantage is that it gives very good conformal coverage of complicated objects, such as drill bits.

Advantages:

- High nucleation density
- Easy scaling to large areas
- Excellent conformal coverage of complex substrates
- Low-temperature deposition demonstrated

Disadvantages:

- Requires conductive substrates
- Low growth rates
- Energetic ion & electron flux at substrate

*Fig. 6.5. Diamond CVD-DC glow discharge.*

Limitations include the fact that the substrate is part of the discharge so that it has to be electrically conductive. Also, the growth rates are abysmally low. Third, this technology cannot be used, for example, for passivating delicate finished integrated circuits because of a multi hundred electron volt ion impact.

We have used DC glow discharge to grow films as thick as 20 microns. These films do remain conducting during deposition. This was surprising at first because it was expected that the electrical conductivity of the film should cease after the first few hundred nanometers of diamond (which is an insulator) are deposited. We do not have proof of what is happening, but these plasma discharges are all accompanied by the production of quite intense ultraviolet light, which I believe switches the diamond film from insulation into photoconduction during diamond growth.

We have grown diamond by DC glow discharge at no higher than 350°C, as determined by eutectic melting specimens. We have no reason to believe that it cannot be done at lower temperatures than that, but we have not made an effort to find out.

The hydrogen content of these diamond films, as measured by Rutherford back-scattering and by forward-scattering analysis techniques, ranges from a low of about one part in $10^4$ to as high as 8 to 10 percent, for a genuine diamond film. For a diamond-like carbon film, the hydrogen content can be as high as 60 percent. These numbers refer to the ratio of atoms. The other principal technique that we use is microwave-excited plasma CVD. In this case, the microwave

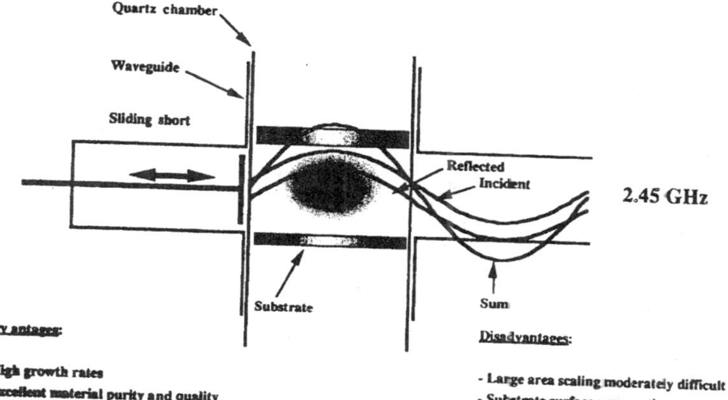


*Fig. 6.6. Microwave plasma CVD.*

generator launches an incident wave across the chamber, where it is reflected from a sliding short that can be adjusted to give superposition of the waves, leading to field intensities high enough to excite ionization in the middle of the chamber.

This technique gives very high growth rates compared to DC glow discharge. The material produced is also of excellent quality in terms of the absence of both non carbon element incorporation and non diamond bonded car bon incorporation. We have used this technique to make polycrystalline films that show better thermal conductivity than the best natural single crystals, and that show similarly good electronic

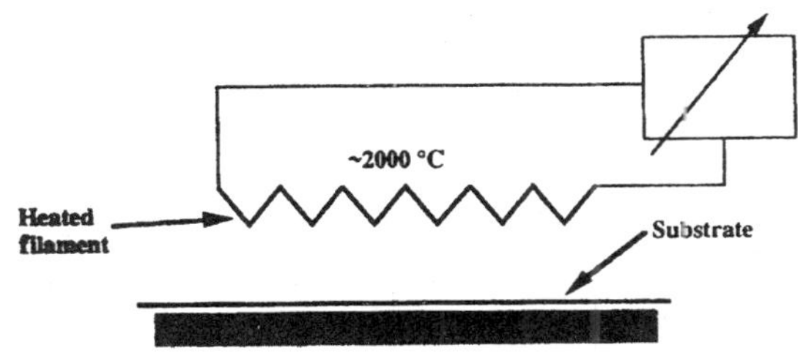


*Fig. 6.7. Diamond CVD-hot filament.*

properties. As already discussed that at 2,000°C, hydrogen and methane react to coat a substrate with a layer of diamond. Go break open a light bulb, put it in a vacuum chamber, and you have all that you need.

The method was a surprise when a Japanese investigator published it, but, in fact, you can use an oxyacetylene torch to make diamond at quite high rates. The irony is that, for as long as people have been trying to synthesize diamond (and modem attempts go back to the last century), it is probable that every apprentice welder (since oxyacetylene welding was invented) has been making diamond while he was learning how to weld. All that is necessary is to have a badly adjusted, fuel-rich flame.

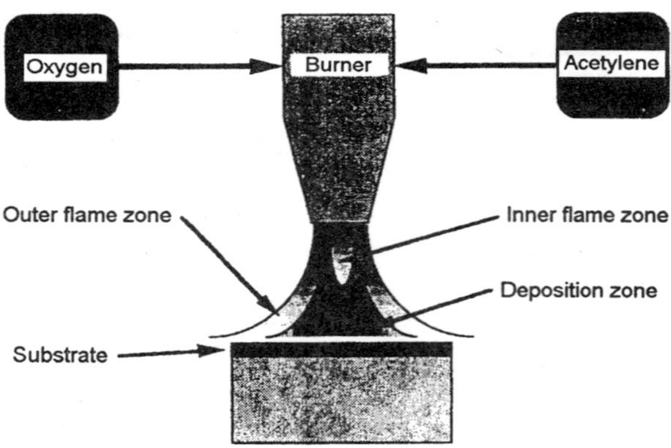


*Fig. 6.8. Diamond combustion synthesis by an acetylene torch.*

A variant on the combustion torch synthesis technique is DC arc discharge and various other plasma torch technologies. The basis of these technologies is a high-temperature arc surrounded with very rapidly flowing hydrogen. Usually the substrate is placed downstream of the arc. At 5,000°K, a supersonically expanding plume from the arc deposits diamond on the substrate. This method gives very, very high growth rates. It is difficult to scale up to cover large areas, but it can be done. An issue of the incorporation of impurities is caused by the erosion of the electrodes in the apparatus, but this issue can be minimized through proper engineering.

All of the techniques described thus far are non equilibrium techniques for diamond synthesis. There do appear to be on the horizon some equilibrium, or purely thermally driven, diamond CVD techniques. This possibility is mostly due to the work of John Margrave at Rice

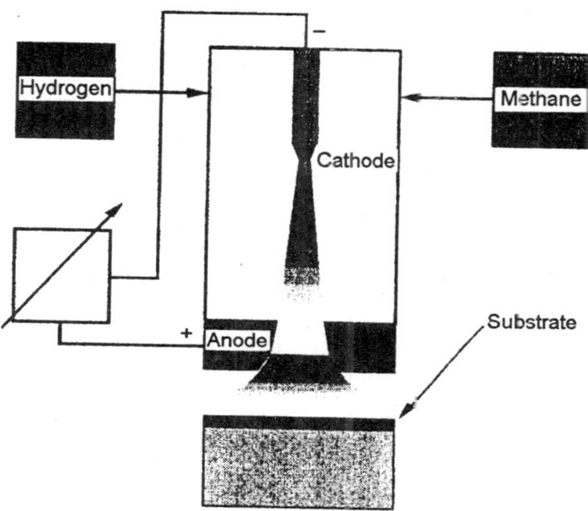


*Fig. 6.9. Diamond synthesis with a plasma torch.*

University. His group has taken fluorine and methane, or $CF_4$ and hydrogen, placed them in a furnace, and behold, they grow diamond! It is not very high-quality material. It has very high dislocation densities, and they have yet to get coalesced films. Nonetheless, it is a demonstration of an approach that I believe has much promise. If this technique can be made to work well, a huge amount of thermally driven CVD equipment in the semiconductor industry can simply be adopted wholesale for growth of this material.

## Applications for Diamond Films

Let us consider the applications for diamond films. These uses demonstrate the huge increments in application performance that are possible by accessing an advanced material—in this case diamond—even in a relatively primitive state of development. We will consider first mechanical applications, specifically cutting tools now and bearings in the near future. In terms of electronic applications, we will consider

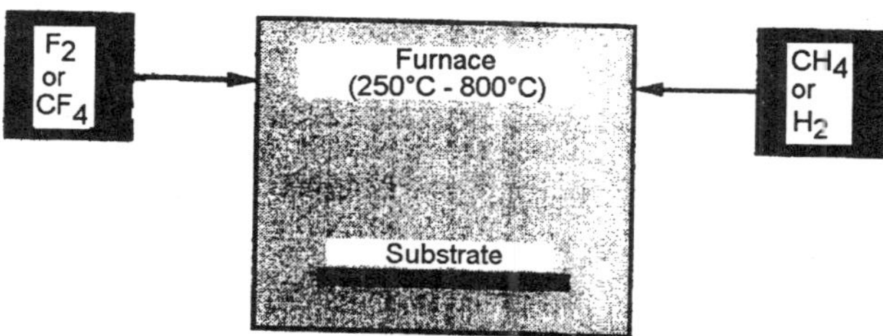


*Fig. 6.10. Diamond synthesis-thermal equilibrium approach.*

device packaging, where the issue is removing the heat from high-power density chips, and future active devices made from diamond. Finally, we will briefly consider some of the biomedical applications.

**Physical Applications**

Various cutting tool geometries are suitable for diamond coating. They all have more or less sharp edges, and are mounted in tool holders that are pressed against spinning pieces of metal.

The performance of some uncoated tungsten carbide tools is compared with the performance of the old diamond technology (polycrystalline diamond compact, or PCD), and with the performance of two diamond-coated tungsten carbide cutting tools. The important thing to note is that the comparison was done under reason ably vigorous machining conditions, and using a material terrible to machine.

Reynolds 390 aluminum is a *hypereutecric* aluminum-silicon alloy containing 18 percent silicon. Hypereutectic means that the aluminum contains more silicon than can be fully dissolved (needed for reasons of mechanical strength), and so some of the silicon precipitates out. This material is industrially important because of fuel efficiency regulations for automobiles. Car manufacturers will have to turn to this alloy in order to meet the weight targets for their cars. It has a really nice strength-to-weight ratio, but it has enough silicate precipitates in it that machining it is like trying to machine a grinding wheel.

Under these conditions, a normal tungsten carbide cutting tool (and this is tungsten carbide with a titanium nitride ion over coating, the best that can be done with that technology) lasts about 20 seconds. It is hard to build a manufacturing line when you have to change tools every 20 seconds. PCD inserts last about nine minutes. The diamond-coated tungsten carbide tools last about the same amount of time, within statistical error. The difference is that the PCD tool costs about $50 and has a labour component of about $35, which is an absolutely irreducible cost. The diamond-coated tungsten carbide tool costs about $5 to start with, and it costs us about $5 to put on the diamond coating. Furthermore, the PCD tool has only one corner and the diamond-coated tools have three corners. The diamond-coated tools, therefore, give a big increment in productivity. Each panel shows a corner of a tool, with the "rake" face on the top and the flat flank along the side. Panel A shows the corner before use, and Panel B shows the corner after machining 390 aluminum. The only effect of wear that can be seen is that the diamond has been polished a bit.

These benefits are the reason that this technology has been developed and others have not. There is an identified commercial demand pulling this technology.

It is necessary to buy 50 carbide inserts to last as long as one CVD diamond-coated insert. At the given selling prices, the customer saves $200 in tool costs, which is a nice saving, but is probably not a sufficient basis to make a purchasing decision. More important is the set-up time, which is the time lost when a tool wears out and must be replaced. At an average cost of $20 per set-up, there is on the order of a $1,000 increment in up-time productivity. Finally, it is generally possible to run diamond-coated tools at twice the surface speed per minute and increase depths of cut, so a huge increase in throughput results. This technology thus saves considerable money for a manufacturer who produces a large number of parts. These numbers were derived from the experience of a manufacturer of aluminum wheels, who produces 1.5 million aluminum wheels per year.

This technology is being extended to more complicated tool geometries—for example, a diamond-coated, tungsten carbide, printed circuit board drill. Current tungsten carbide bits used to drill fiber-reinforced printed circuit board materials give about 1,200 hits, which is a tool life time of only 10 minutes. Typically, a couple hundred drills are used in a manufacturing operation, so it is important to extend the wear life of the drill bits. A thirty fold to fifty fold increase can be achieved with a 10-micrometer coating, but the reproducibility is still low. Part of the problem is the fact that the thin printed circuit board drills are slammed into the boards at such velocity that they are torquing considerably. The difficulty is to make the diamond coating bond well enough to the drill bit that it does not break loose and peel off.

A more recent concern is diamond-coating of bearings. A plot showing the relationship between the log of Knoop Hardness Scale and temperature for a number of different materials reveals that the curve shown for diamond (which is for natural diamond crystals) is very high. The steels used in bearings are centered in the bottom part of such a diagram. Material hardness is one of the principal predictors of bearing life. Therefore, diamond should do very well for us in applications to advanced bearings that last a long time.

Such diamond bearings would not require lubrication in ordinary operation because the hydrogen termination (of the unsatisfied valence bonds of carbon atoms on the surface of the crystal), which is principally responsible for the low sliding coefficient of friction, remains on the surface of the diamond in air. We have done some work already with diamond coating of silicon nitride bearings. This material is easy to coat. Silicon nitride bearings are used in high-end gyroscopes and in turbo pumps for high-vacuum systems, where clean and lubrication-free operation is mandatory.

**Advanced Computerized Applications**

A major use of diamond in electronic applications is for heat dissipation. The best ceramics, aluminum nitride and beryllium oxide, are about 2 to 3 watts per cm per degree, depending on how they are processed. Natural diamond and CVD diamond (with about tenfold higher conductivity), obviously offer good potential. Recently, CVD diamond has even slightly exceeded natural diamond in conductivity.

The impact of this property of diamond upon device performance can be seen from a multichip module package thermal simulation out of MCC (Microelectronics and Computer Corp., a government-sponsored research consortium) done by Nalan Kumar. The results indicate the total amount of power that can be put into their reference chip set, if the chip set is packaged in plastic versus in various other materials. Given that an electrical insulator is required, the very best that can be done with non diamond technology is just more than a third of what can be done with diamond. Operational results from laser diode arrays using diamond heat sinks provide experimental proof of these

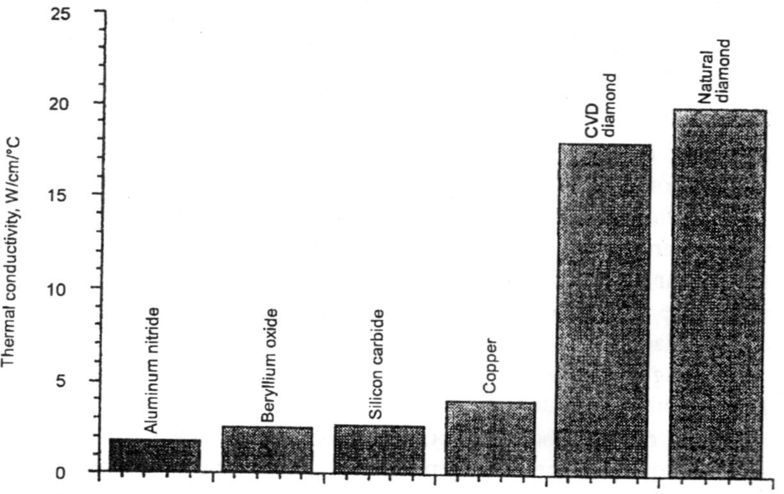


*Fig. 6.11. Thermal conductivity of selected electronic packaging materials.*

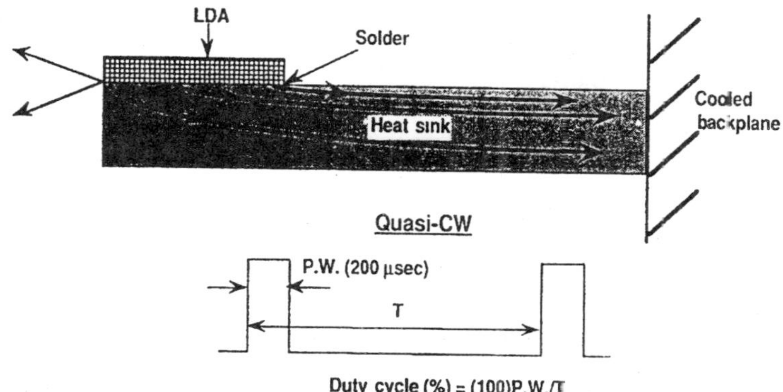


*Fig. 6.12. Laser diode array mounting configuration on a heat sink.*

expectations. An array of laser diodes (which is manufactured in a monolithic gallium arsenide bar) is shown in the upper-left corner, extending perpendicular to the plane of the figure. The light emitted from the diodes extends to the left from the junction between the diode and the heat sink. The laser diode is eutectically bonded to the heat sink. The bonding with diamond is achieved through a three-layered metallization sequence. The first layer is titanium, which forms some sort of a titanium-carbide with the diamond. After that comes a platinum layer, and on top of that a gold surface, which then can be used for eutectic bonding. The large arrows in the figure indicate the flow of heat through the heat sink to the cooled back plate. The power profile put into the laser diode array is diagrammed below the schematic of the diode array and heat sink.

The upper trace shows the result if the laser diode array is mounted on a copper heat sink, while the lower trace shows the result of mounting the laser diode array on a piece of diamond of equivalent geometry. Copper is the best of the metallic conductors. Nevertheless, the junction temperature using copper limbs within a few milliseconds to an asymptotic temperature of about 41°C, whereas with diamond it stays at about 31°C to 32°C. This difference in thermal conductivity means a large difference in power output, because these devices are incredibly temperature-sensitive. Our customer mounted their laser diode array on diamond in exactly the same way that they mounted it on copper. They got about 2 watts more optical power out, solely as the result of increased junction efficiency caused by lower temperature operation. The variation of electrical conductivity with temperature of CVD diamond film compared to natural diamond. CVD diamond (the

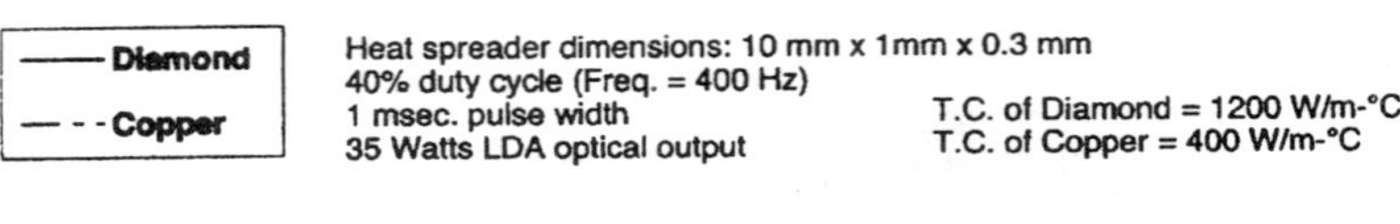


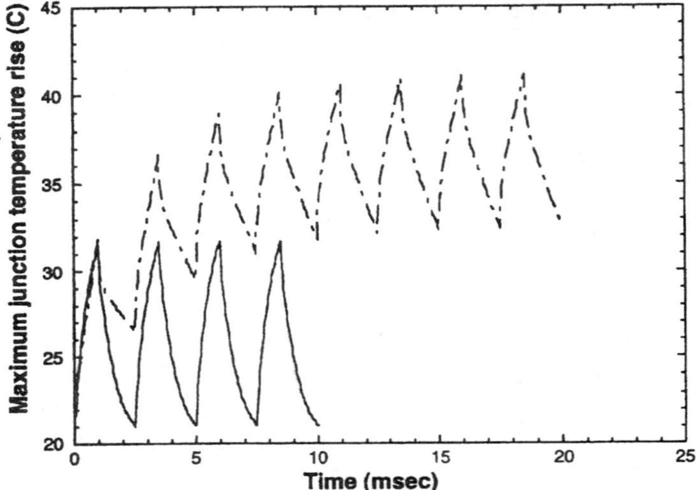


*Fig. 6.13. Simulation of the junction temperature of a laser diode array.*

triangles) is two orders of magnitude better in terms of electrical resistivity at elevated temperature.

In terms of mobility versus carrier density, generated with ultraviolet excimer laser excitation. The performance of the film made

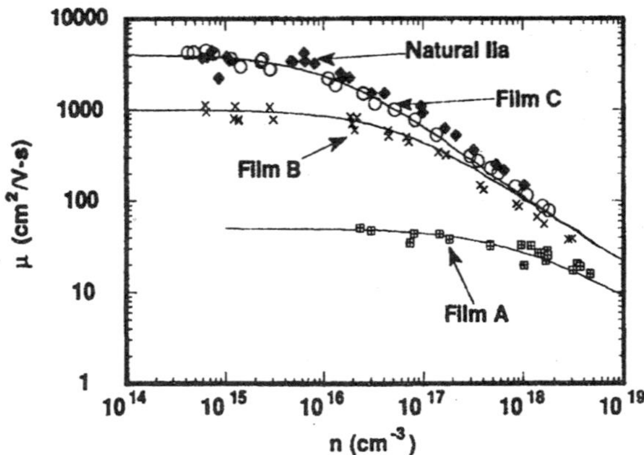


*Fig. 6.14. The low-yield mobility of CVD diamond film equals that of natural IIa single-crystal diamond.*

in 1992 is equaling that of natural single-crystal diamond. It took nature a billion years to get to carrier collection distances of 20 to 50 microns. That is an index of how far a carrier travels before it is collected. It has taken the industry about three years to cover the same ground, so there is good potential there. In other applications, we are making diamond photoconductors into photo detectors. These are fairly crude devices, but they still perform well. We are also making silicon on insulator structures, where the insulator is diamond. That has, in addition to the usual radiation hardness benefits that you get from silicon on sapphire, the enhanced thermal conductivity of the diamond.

Finally, we know how to etch diamond. Currently, a feature size of about 10 micrometers seems achievable, which is comparable to the grain size. As a resist, both gold and $SiO_2$ films have been employed. Etching can be accomplished in an oxygen plasma at the rate of about one micrometer per minute.

**Biomedical Applications**

There are a number of promising potential biomedical applications for CVD diamond. The applications make use of one or more of diamond's unique material properties. For example, diamond's extreme hardness, in combination with its chemical stability, allow it to be used as scalpels for surgical procedures. The properties of diamond scalpels include more rapid wound healing, because a minimum of collateral tissue damage occurs when diamond blades are used. Currently, single-crystal diamond gems are used as scalpels in ocular surgery. When processes are developed to allow diamond coatings to be applied to steel, diamond scalpels will be affordable for more common surgical procedures.

A second area of application for CVD diamond coatings is as a component of prosthetic joint implants. Here, diamond's hardness, inherent lubricity, and low immunoreactivity will combine to extend the operational life of implanted artificial joints, reducing the frequency of replacement procedures, which are painful and disabling.

## Modern Synthesis Technologies

The diamond synthesis technologies that we are using now and in the near future are all "*shake-and-bake*" chemistry. Until we have nanotechnology functioning. Nevertheless, we can expect some improvements in this "*shake-and- bake*" technology in the near future. One improvement will be fluidized bed CVD—that is, plasma discharge CVD operated in a bed of particles to grow cheap diamond particles for diamond ceramics.

Another improvement will be in *liquid-phase systems*. These are principally liquid metals. In these systems, it is possible to get good control over atomic hydrogen-activity levels, and good control over carbon activity levels. This control should allow us to grow high-quality single-crystal materials, and also whiskers of diamond. Whiskers of 100-micrometer length would already be useful for fiber-reinforced ceramics.

*Molecular beam chemistry* will probably also be important in the near future. It is now possible to make clean beams of methyl radicals and atomic hydrogen, two species that do work for making diamond. The issues here are beam brightness and beam area. The main reason for pursuing this area would be diamond electronics. But useful, high quality transistors are still maybe three to five years off. Finally, in terms of what we will be able to do in the more distant future, having assemblers capable of making diamond fibers will make possible the tower reaching to earth orbit and the elevator to geosynchronous earth orbit, subjects familiar to science fiction fans.

# 7

# Environmental Regulation of Nanotechnology

The relationship between new technologies and the environment is a complex one. On the one hand, various human technologies—ranging from "low" technologies like slash-and-burn agriculture, to "high" technologies like nuclear weapons—have done more than their share of environmental harm. On the other hand, new technologies are often cleaner and safer than the older technologies they replace, and may offer ways of remedying environmental harms previously thought of as beyond help.

Both of these aspects are likely to come into play with molecular nanotechnology, a technology so new that, in truth, it barely exists yet. But though the actual accomplishments of nanotechnology at this date fall into the workbench or proof-of-concept stage, research is progressing at a speed that outpaces the predictions of the most optimistic prognosticators. (Indeed, nanotechnology has received so much attention—not all of it positive—that some are already pronouncing it a cliche.) If researchers continue to make progress at this rate, nanotechnology will hit the marketplace more quickly than did biotechnology, a field of endeavor to which society is still adjusting. It thus seems worthwhile to begin the discussion now.

This all-too-brief essay will outline the basic nature of molecular nanotechnology. It will then discuss the likely environmental benefits (environmentalist Terence McKenna, writing in the *Whole Earth Review*, called *nanotechnology* "the most radical of the green visions") and harms (some critics worry that rogue nanodevices will devour the planet) of this technology, and at least seek to begin the discussion of how nanotechnology might be dealt with in a way that will maximize the environmental benefits—which are likely to be enormous—while minimizing the potential harms, which, if allowed to materialize, are likely to be large as well.

## Science and Technology of Nanotechnology

### How Nanotechnology Works

Put simply, nanotechnology is a technology for making things by placing atoms precisely where they are supposed to go. Traditional industrial technologies operate from the top down. Blocks or chunks of raw material are cast, sawed, or machined into precisely formed products by removing unwanted matter. Results of such processes may be rather small (integrated circuits with structures measured in microns, for example) or very large (ocean liners or jumbo jets). However, in all cases matter is being processed in chunks far larger than molecular scale.

We are used to this kind of top-down technology, and it is certainly capable of yielding products of fairly high precision and complexity. It is the basis of our civilization, and it has brought us the many technological revolutions described above. It is, however, something of an aberration in the natural order of things, as most products of living organisms—and those organisms themselves—are made very differently.

Rather than being produced through large chunks of material being sawed, planed, and ground to form, most such objects are constructed by tiny molecular machines, such as cells and organelles, working from the bottom up. By organizing individual atoms and molecules into particular configurations, these molecular machines are able to create works of astonishing complexity and size, such as the human brain, a coral reef, or a redwood tree. This approach can produce results that would seem impossible if judged by the standards of conventional top-down production technology, but that are taken for granted in their proper context. For example, the human body begins as a single cell, a fertilized ovum. Yet a mature human being consists of approximately 75 trillion cells, complexly arranged and of many different varieties. The molecular machinery responsible for this amazing, though commonplace, feat of production is capable of such dramatic results because it performs operations in parallel (that is, with many cells operating at the same time through most of the growth process), and from the bottom up.

As Eric Drexler states: Nature shows that molecules can serve as machines because living things work by means of such machinery. Enzymes are molecular machines that make, break, and rearrange the bonds holding other molecules together. Muscles are driven by molecular machines that haul fibers past one another. DNA serves as a data-storage system, transmitting digital instructions to molecular machines, the ribosomes, that manufacture protein molecules. And these protein molecules, in turn, make up most of the molecular machinery just described.

Putting these natural molecular machines to work is nothing new, of course, as every living thing does so constantly. Nor is deliberate human programming of those machines particularly new, as it is what genetic engineering (or even selective breeding) is all about. What makes nanotechnology different is that it involves the attempt to go farther than natural mechanisms permit. Using special bacterium-sized "assembler" devices, nanotechnology would permit exact control of molecular structures that are not readily manipulable by organic means (diamond, or heavy metals, for example) on a programmable basis.

With nanotechnology, atoms will be specifically placed and connected, all at very rapid rates, in a fashion similar to processes found in living organisms. Trees, mammals, and far less complex organisms make use of molecular machinery to manufacture and undertake repairs at a cellular and subcellular level. The key to the application of nanotechnology will be the development of processes that control placement of individual atoms to form products of great complexity at extremely small scale.

This approach was originally suggested by physicist Richard Feynman. In an article entitled *There's Plenty of Room at the Bottom*, Feynman explored the potential of atomic-scale physical manipulation of matter. As Feynman said: The principles of physics, as far as I can see, do not speak against the possibility of maneuvering things atom by atom. It would be, in principle, possible ... for a physicist to synthesize any chemical substance that the chemist writes down.... How? Put the atoms down where the chemist says, and so you make the substance. The problems of chemistry and biology can be greatly helped if our ability to see what we are doing, and to do things on an atomic level, is ultimately developed—a development which I think cannot be avoided.

Scientists and researchers are making progress in this direction today. Already, IBM's research division has demonstrated the ability to manipulate individual atoms by constructing a copy of IBM's logo out of individual xenon atoms, manipulated by the tip of an atomic force microscope. The next step, the precise placement of atoms in combination to form stable compounds, and structures, has also been achieved.

Such efforts have already generated a substantial amount of theoretical literature, and considerable concrete interest. Nanotechnology has already produced a number of books and articles, government reports, and at least one well-established and well-funded research program—unfortunately in Japan, not the United States, though the United States is now forging ahead with its own National Nanotechnology Initiative.

**What Nanotechnology Can Do**

Full-fledged nanotechnology promises nothing less than complete control over the physical structure of matter—the same kind of control over the molecular and structural makeup of physical objects that a word processor provides over the form and content of a text. The implications of such capabilities are significant: to dramatize only slightly, they are comparable to producing a 747 airplane or an ocean liner from the mechanical equivalent of a single fertilized egg.

Using nanotechnology, production would be carried out by large numbers of tiny devices, operating in parallel, in a fashion similar to the molecular machinery already found in living organisms. However, these "*nanodevices*" would not suffer from the constraints facing living organisms—they would not have to be made of protein, or other substances readily extractable from the natural environment, nor would they have to be capable of reproducing themselves. Instead, they could be constructed of whatever material, and in whatever fashion, is most suited to their task. Known as "*assemblers*," these tiny devices would be capable of manipulating individual molecules very rapidly and precisely. The process of using such assemblers to manufacture products may be hard for many readers to visualize; the following explains how this could work.

Nowadays, some medicines are made through biotechnological processes, for example those using recombinant deoxyribonucleic acid (DNA). In essence, this means that the DNA of living creatures (usually bacteria) is altered so that the creatures are reprogrammed to produce the desired substance by assembling component atoms into the desired configurations: hydrogen here, carbon there, and so on. This approach represents a revolution in pharmaceutical technology, but has distinct limitations. Since biotechnology is based on altering the program of living organisms, only substances that can be handled by living organisms can be manufactured; only mechanisms possessed by living organisms can be used. It is as if clothing were manufactured by training spiders and silkworms to weave their product in particular patterns.

By contrast, modern textile technology represents a far more powerful, more versatile, and easier approach to manufacturing clothing. Nanotechnology represents a similar approach to the manufacture of other goods, including pharmaceuticals. Imagine the power and complexity of today's computer-driven textile looms put into machines orders of magnitude smaller than the period at the end of this sentence. Instead of weaving cloth, such machines would seize individual atoms using selectively sticky manipulator arms, then "plug" those atoms together (somewhat like assembling "lego" blocks) until chemical bonding took place. By repeating these steps according to a programmed set of instructions, a nanotechnological approach would be able to produce substances that conventional biotechnology could not (say, because they are toxic to living organisms, or use elements that living organisms cannot handle efficiently) and would be able to do so with greater speed and lower expense. This advantage would increase with an increase in complexity on the part of the desired molecules.

With relatively mature technology, we might expect to see general-purpose chemical synthesizers using nanotechnology. The desired molecule would be modeled on a computer screen, the assemblers would be provided with the proper feedstock solutions, and the product would be available in minutes. This application of nanotechnology would be relatively simple. More complex applications might use groups of assemblers programmed to produce molecules and then hook them together into large structures: rocket engines, computer chips, or whatever is desired.

Besides allowing such efficient and powerful manufacturing capabilities, more sophisticated applications of nanotechnology would allow far more subtle applications. For example, specially designed nanodevices, the size of bacteria, might be programmed to destroy arterial plaque, or cancer cells, or to repair cellular damage caused by aging, and then be injected into the body. After performing their tasks, the devices may be induced to self-destruct, or remain in a surveillance mode, or, in some cases, integrate themselves into the body's cells. Such devices would have dramatic implications for the practice of medicine, and for society as a whole.

## Environmental Risks and Benefits

Obviously, a technology of such capabilities offers both upsides and downsides. Some are beyond the scope of this Article: for example, Bill Joy's celebrated fear that nanotechnology may lead to super-intelligent machines that will use their intelligence to achieve world domination and replace humanity. (Personally, I find this unlikely. A simple glance at the headlines should be enough to dispel the belief that world domination is secured through superior intelligence. At least, if that happens, it will be the first time in human history that it has occurred that way.)

### Risks

There are, however, genuine environmental risks to nanotechnology, and the nature and extent of these risks has occupied the literature in the field since the beginning. At the grossest level, there is the fear that someone will design self-replicating nanorobots, capable of making copies of themselves from materials found in nature, and that those nanorobots will convert everything in the world into copies of themselves, thus wiping out the entire biosphere. In nanotechnology circles, this is known as the "*gray goo*" problem, after the notion that such uncontrolled replication will lead to the entire world being turned into, well, gray goo.

Such a notion is, to say the least, disturbing. While it is probably true that we already possess the capacity to substantially destroy the biosphere using nuclear weapons, or even advanced military biotechnology, the prospect of an unstoppable wave of nanorobots devouring everything in a fashion vaguely reminiscent of the science fiction movie *The Blob* seems somehow spookier and more frightening. Fortunately, further study suggests that such an event could take place only by deliberate action, not by accident.

Deliberate action, however, is not impossible. More serious environmental threats involve the military or terroristic use of nanotechnology (already, the U.S. Department of Defense is one of the main sponsors of nanotechnology research). Military nanotechnology is likely to be less grossly destructive than self-replicating "gray goo," but harmful enough in its own way. Early military uses of nanotechnology are likely to involve sensors and similar comparatively benign applications. More advanced applications, however, are likely to be both powerful and subtle: devices that can infiltrate electronics and seize control at crucial moments, artificial "disease" agents that can rest harmlessly in victims' bodies until activated by an external signal, and so on. Some of these will have no significant impact on the nonhuman environment, but nanotechnology-based agents for crop destruction, forest-cover removal, and area-denial applications are likely to pose familiar environmental problems in a new fashion. On the more positive side, it is possible that such agents will be less persistent and less broadly destructive than, say, Agent Orange or conventional land mines. Set against this possibility is the prospect that these agents will be used more broadly for that very reason, along with the possibility—impossible at this point to quantify, but irresponsible to dismiss—that they will not work as well or as "safely" as intended.

Nanotechnological devices for military use also raise the issue that they do the work of chemical and biological weapons, but—at least arguably—do not fall within treaties regulating chemical and biological weapons. The argument that nanotechnological weapons—at least those of destructive, rather than surveillance, type—would be functional equivalents of chemical and biological weapons would be a strong one, and indeed destructive nanoweapons would probably achieve their effects through chemical action, though it would be mechanically initiated. Nonetheless, as the Reagan Administration's efforts to reinterpret the ABM Treaty illustrate, national governments do not require much encouragement to advance novel or disingenuous interpretations of the law where doing so serves their interests.

Controlling these military applications will be difficult. Military interest in nanotechnology is already high, and an unknown, but large, amount of military nanotechnology research is going on at present; this is sure to increase as the actual application of nanotechnology becomes more feasible. It is not too early, however, to look at updating the chemical and biological warfare conventions, and other related instruments, and to explore ways in which employment of destructive nanotechnology is constrained by the laws of war.

There is also a risk that civilian applications of nanotechnology will not work as well or as cleanly as expected. This risk is almost certainly smaller than similar risks associated with military technologies, since civilian technologies tend to be more robust, and founded on a much deeper experience base than military technologies. (This characteristic is even more pronounced when the manufacturing or coding standards are open, which is why "open source" software is generally more reliable and robust than proprietary "closed source" software. This lesson has not been lost on nanotechnology enthusiasts.) Nonetheless, as nanotechnology moves out of the laboratory and into the marketplace, it will be important to develop standards that ensure that its products (and by-products, if any) do not have dangerous effects.

In many ways, these risks are likely to be lower and more manageable than those associated with biotechnology, both at military and civilian levels. Although one can imagine nanodevices of fiendish subtlety and destructiveness, it will in fact be quite some time, if ever, before it is possible to design a nanodevice that approaches the smallpox virus (even in its pristine, non-biowar form) for virulence or lethality. Despite fears of self-replicating nanodevices, the world is already full of self-replicating lethal agents that menace us on a continuous basis, and governments have been working for decades to make those agents more lethal. Most of our problems in coming decades will stem from these agents, not from conjectural nanoplagues.

**Benefits**

If the environmental dangers of misapplied nanotechnology are significant, the environmental benefits of nanotechnology properly employed are dramatic; it was not hyperbole when Terence McKenna called nanotechnology a *radical green vision*. Since nanotechnology involves atom-by-atom construction, it will be able to create substances, and even finished objects, without producing the dangerous and messy by-products that most current manufacturing processes produce. Nanodevices will operate in a liquid containing the necessary raw materials (usually carbon or silicon, with trace amounts of other elements as needed) and will simply plug the appropriate atoms in the appropriate places to produce the desired end product. Such processes should produce few by-products, and those by-products can be readily purified (by other nanodevices) and recycled back into feedstocks.

What is more, most products of nanotechnology will be made of simple and abundant elements: carbon, in diamond or diamondoid form, is seen as the basis of most nanomanufacturing. Products made of such materials will be very strong, meaning that smaller amounts of material can be used, and carbon is an abundant material, meaning that little in the way of exploration and extraction will be needed. Indeed, as a greenhouse remediation measure, nanodevices could even extract carbon dioxide from the air if desired.

This clean manufacturing is a significant benefit of nanotechnology, but in some sense it is less important than the economic regime that it makes possible. When materials are inexpensive, and structures of great strength and low weight can be manufactured cheaply, energy requirements for many activities drop enormously. If, for example, we can make cars that are stronger and safer than contemporary vehicles, but that weigh one-fourth (or even one-tenth) as much, electric vehicles become far more practical. Indeed, with materials of very high strength to weight, solar-powered vehicles become practical. Similarly, strong, inexpensive, energy-efficient buildings would become far more practical, further reducing energy demand. Many experts also believe that atom-by-atom manufacturing will make low-cost, high-efficiency solar cells practical. The combination of reduced energy demand and inexpensive solar power may make most of today's power generation and transmission infrastructure unnecessary. The payoff from such improvements would be enormous, not only in terms of reduced pollution from power generation, but in terms of reduced environmental impact all along the production and distribution chain: less damage from power-line construction and maintenance, less damage from transformer leakage, and, of course, less damage from coal mining, oil extraction, nuclear fuel-cycle operations, and so on.

Beyond these impacts, more advanced nanotechnology may allow active remediation of many environmental problems. For example, toxic wastes in contaminated aquifers may be neutralized by specially designed nanorobots that selectively capture undesirable molecules and then either sequester them for removal or (where the danger is chemical, not nuclear) break them down into harmless substances. While nanodevices cannot, for example, render radioactive materials nonradioactive, they could capture molecules of radioactive waste and concentrate them into a form that would be easily removed.

## What to Do

At this early date, nanotechnology remains mostly a matter for laboratory work and computer simulations. That, however, makes this a good time to think about ways of regulating nanotechnology that will allow us to reap its benefits while avoiding the harms that can result from misuse. In 1999, the Foresight Institute, a Silicon Valley foundation devoted to exploring issues relating to advanced technologies, sponsored a conference intended to start the discussion on regulation of nanotechnology. Conference participants included representatives from the scientific, industrial, environmental, academic, and defense communities, and comprised a very diverse group of individuals, backgrounds, and agendas. Surprisingly, there were several areas of consensus.

The first was that simply renouncing nanotechnology—even if such a course were deemed desirable—is impossible. Unlike nuclear weapons research (which itself is poorly controlled), nanotechnology research and development does not require a large or specialized infrastructure. Indeed, the conceptual work has largely been done, and although many technical difficulties remain to be overcome, the barriers to the implementation of nanotechnology are in the nature of engineering, not basic science. In short, it was agreed that nanotechnology would be developed regardless of any efforts to suppress it, and that such efforts would only ensure that whatever research took place would do so in rogue nations, with few constraints.

The second was that proper regulation, and the use of what Arthur Kantrowitz has called "the weapon of openness," could nonetheless serve to control risks of improper use of nanotechnology. The Foresight conference produced a set of draft guidelines for research and use of nanotechnology that were designed to achieve these ends, and this draft was placed on the Foresight website for comment and critique. Many comments and criticisms were incorporated in several revisions of the draft, the current version of which is attached as an appendix to this Article. This process continues, and readers are encouraged to participate.

The third conclusion was that regulating nanotechnology will be a process, not an event. While it is not too early for thought, it is certainly too early for legislation, and an attempt to create some sort of overarching legal code for nanotechnology in advance of the facts would likely be disastrous in light of our imperfect knowledge at this point. As technology develops, and as society changes, regulatory approaches will have to keep pace. We are fortunate that we have some years—or perhaps a couple of decades—before such matters become urgent. We would be wise to make use of that time to think things through, before events surpass us.

## CONCLUSION

This all-too-brief discussion has outlined the basic character of nanotechnology and the risks and benefits it presents. The technology of the very small poses issues sufficiently large that many minds will be required to address them. He encourage readers to add their thoughts to the process. More than most lawyers, environmental lawyers have firsthand experience with the considerations (and limitations) inherent in the application of law to advanced technologies amid conditions of technical and societal uncertainty. That expertise is likely to be helpful.

# 8

# HOMELAND NANOTECHNOLOGY

In the past, cities were walled and fortified to withstand attack from the outside. Settlement sites were selected not just for industrial purposes or for access to transportation as they are now, but because they were defensible. In today's world of bombers, missiles, tanks, and other mechanized combat equipment, the idea of fortifying a city seems to be obsolete. But this is not the case. Though walls may no longer do the job, modern defenses and security measures exist to harden critical infrastructure and lessen vulnerability to terrorist attacks.

Buildings are often engineered to withstand earthquakes and other predictable disasters (the World Trade Center towers were actually designed to withstand the impact of a small aircraft), but the issues of bomb resistance, air filtration, and threat detection are only modest concerns in the planning of most cities and structures. Excepting the construction of fall-out shelters, hardening against the threat of nuclear strikes has been considered totally outlandish and not worth consideration. This perception must change. In a world tainted by terrorism, the buildings likely to be targeted can and should participate in their own defense.

Although major office buildings have been the targets of the most deadly terrorist attacks in America, more critical infra-structure often goes virtually unprotected. The minimal monitoring of water stations, mail sorting offices, power distribution points, and even vital telecommunications facilities make them highly vulnerable to chemical or biological attack. Despite the government's emphasis on airport security, airports are not much better, and bridges and train stations have essentially no security of any kind. Finally, the people required to respond first if any of these places are threatened—firefighters, police, and paramedics—need better protection on the job. Lives depend on their having it.

As always, the best way to defend against attack is to prevent the attack from occurring. The FBI and other intelligence agencies have been doing an impressive job since 9/11 of predicting attacks and finding culprits, but they need tools that allow them to extract information without unreasonable intrusion innocent people's lives. These tools include better technology for detecting qualified threats and more effective decryption for intercepting communications. These technologies will require oversight, but that is no reason to delay their development.

Terrorist attacks are the kind of disaster that America now dreads most keenly. Unfortunately, it seems likely that another major attack will occur on American soil. Nanotechnology will help mitigate the damage caused by such an attack, though only changes in policy that are beyond the scope of this book can eliminate the threat entirely. Other types of crises unrelated to terrorism could also be addressed by nanotechnology. Fires, floods, natural disasters, and the outbreak of disease are all areas where nano-applications can make a significant positive difference. So nanotechnology for homeland security may have even more dual-use applications than nanotechnology for military applications and it is easy to see how it could directly impact all of our lives.

## Hardening the Hearts of Cities

In recent years, the most harmful terrorist attacks on America have been explosive attacks on large buildings. The first World Trade Center bombing, the Oklahoma City Federal Building bombing, and the 9/11 attacks all involved the destruction of office complexes. Commercial landlords and civic authorities reacted to the 9/11 attacks by implementing cursory inspections for vehicles pulling into parking lots under high rise buildings and by imposing more vigorous campaigns to check visitors' photo identification. While these precautions may help, how can parking attendants hope to find bombs that may be concealed in a suitcase or even under the seats of an SUV? How many would recognize a bomb even if they found it? And what prevents a terrorist from forging a photo ID?

Some of the sensor technology could be applied to inspecting vehicles, but a bomb can also be detonated from the street in front of a building and a car containing a bomb may run into or through a barricade. Individually harmless chemicals can be brought into buildings on separate trips, then mixed therein to produce disastrous weapons. How can office and apartment buildings be hardened to reduce the impact of such attacks? Nanotechnology offers several solutions that could be integrated into the design of future, attack-resistant buildings.

Start by considering structural issues. In San Francisco and other key earthquake zones, buildings are designed to be strong but also flexible. Some designs utilize a flexible central mast to support the rest of the building. Constructing these masts of nano composite materials or using nanocomposites to replace today's ubiquitous structural steel and reinforced concrete could significantly improve buildings' performance and protection.

The structure of a nanocomposite material resembles that of a normal composite material such as reinforced concrete, but shrunk to the nanoscale. Reinforced concrete consists of ordinary concrete poured over a steel mesh called *rebar*. It is formed at the macroscale, but it still manages to combine the hardness and compressive strength of concrete with. The yield strength of the rebar to create a material with properties superior to either of its components alone. Nano-composites could work the same way, but at the molecular level. For example, a nanocomposite might be made of a high-strength plastic wrapped around a rebar of nanotubes. Such a material could vastly outperform conventional construction materials, and a building based on it could withstand a much stronger explosion than anything that exists today.

Explosions typically cause damage in two ways. The first is through the blast concussion. The second is through heat, which is what actually caused the collapse of the World Trade Center towers. (The impact of airplanes alone would not have toppled the towers, but the immense heat generated by the burning jet fuel weakened key girders and supports.) Advanced nano materials can be used to deal with this problem. Buildings incorporating this and other nano materials now under development will have a much greater chance of surviving the heat of explosions and other kinds of fires.

Because of their immense strength, the amount of these nano materials required to build a nano-enhanced building is actually much smaller than the amount of conventional materials that would otherwise be used. This opens up whole new design possibilities and the nano-school of architecture may far be far away.

## Smelling Smoke

Buildings and homes are already equipped with sensors for many common hazards. Almost all cities have codes requiring *smoke alarms* and *carbon monoxide detectors*. Even *burglar alarms* are a kind, of sensor, though for a very different sort of problem. Continuous environmental monitoring for chemical and biological weapons may never make sense for individual homes (though concerned citizens will certainly have the option to install such systems), but it makes a great deal of sense to put sensors in the air and water handling systems of large office and apartment buildings.

Most modern high-rise buildings already contain complex air handling systems for central air conditioning, heating, and humidity control. In many of these buildings windows cannot be opened for climate-control efficiency and safety reasons, but that makes them more vulnerable to attack. If a toxin were introduced into the air supply, getting clean air would require breaking the windows (causing another hazard in the form of falling glass), and clearing the building afterward would be extremely difficult.

If *nanosensors* were installed in high-rise air handlers and if the same kinds of filters we looked at for next-generation gas masks were introduced into the systems, the possibility for contamination would be reduced significantly A building could be evacuated if a toxin were detected and air could be safely filtered in most cases. During day-to-day operations, these detectors and filters would be low cost and low maintenance. They might also have the beneficial side effect of extracting ozone, particulates, and other common pollutants from the building's air. In this case, preparing for disaster would also mean improving air quality all of the time.

These same kinds of hardening processes could also be implemented for key infrastructure like water, electricity, sub ways, mail centers, and communications. In these cases, the upgrades are not really optional. To fail to equip a water pumping and filtration station with state of the art sensor technology, for example, would be simply negligent. Remediation in the event of disaster would also have to begin at these points, so it makes sense to keep a supply of common remediation compounds on hand.

## First Response

The term *first responders* has grown to mean police, firefighters, paramedics, and all other emergency workers who represent the front line of defense in case of a terrorist attack within the United States. These men and women face above-average risks in their day-to-day work, and the additional threat posed by terrorists both exacerbates certain basic problems and introduces new ones.

Like soldiers, police officers are at risk from enemy fire and often have to wear cumbersome protection including bulletproof vests while performing their duties. Firefighters require full sets of fire retardant suits, portable oxygen supplies, and axes, but despite these precautions even the best-protected firefighter can only remain in a burning building for a few minutes. Paramedics routinely deal with sick and contagious people, which is essentially a lower level biological attack. What can nanotechnology do to help?

Clearly the advanced-armor nanotechnology for soldiers and heat-resistant nanotechnology for buildings will help first responders. Chemical-defense nanotechnologies will be useful not only for their stated purpose, but also for use in filtering smoke and various toxic chemicals from building fires. Nanotechnology sensors may also be useful in replacing canine units for many purposes—even a dog's impressive sense of smell cannot detect a single molecule of something. Sensors will certainly have applications for diagnosing people and determining the threat level of a situation.

Nanotechnology also has some specialty applications for first responders. When such professionals discover a bomb, the protocol is to remove and safely detonate it if possible. This may require transporting a device and taking some risk of detonating it by mistake. To help reduce the impact of such an accidental explosion, nano materials like US Global Nanospace's Blast-X can be integrated into wall panels to significantly reduce the effect of an explosion. Such materials might also be used as blast retardants in trucks, carrying cases, and even specially modified body armor.

First responders also need remediation agents (chemicals that can break down nasty molecules or biotoxins) at least as much as the military does. One interesting solution to the problem involves using photo catalytic self-cleaning nanolayers (sometimes called *PSC layers*). These are surface coatings consisting of a nano-thin layer of material (sometimes something as simple as silver titania) that can break down various harmful contaminants when exposed to light. The National Technology Transfer Center's Emergency Response 'technology Pro gram claims that anthrax, smallpox, botulinum toxins, ebola, cholera, bubonic plague, nerve agents, mustard agents, hydrogen cyanide, tear gases, and even exhaust fumes and smoke can be broken down by exposure to PSC layers. These nanolayers could find applications in chemical-defense masks, air handling systems, and self-cleaning lighting, paint, glass, and other household fixtures. PSC technology has been developed by PPG, GE Lighting, Pilkington Glass and a few groups in Japan, where the idea has been championed by Akira Fujishima of the University of Tokyo.

There are other promising technologies for first responders. Many of them are very similar to the soldier nanotechnologies we have already discussed (although missing the active camouflage and weapons kit).

## Clean It Up!

Decontamination following either an attack or an accident is a significant challenge in homeland defense. If prevention fails, first responders can deal with the immediate issues of safety and containment, but the threats of dispersed toxic chemical or biological agents require broad remediation and decontamination strategies.

We have already noted that chemical reactions occur at surfaces, and that *nanoparticles* have huge ratios of surface to volume because of their tiny size (half an ounce of nano-powdered alumina has more surface area than a football field). So *nanocrystals* and powders will react with toxic agents far more rapidly and effectively than traditional crystals or powders. This suggests that decontamination with such powders can be rapid and complete. The powders can either adsorb the toxic species (much like World War I gas masks that used activated carbon) or react with and chemically destroy the toxin (like the PSC layers discussed above). The powders can actually he sprayed on or applied with special mittens being developed by the Army Research Laboratory.

Decontamination of equipment, buildings, and even people is important enough that it already has a name (Decon) and some component serial numbers (M-11 and M-295) in the U.S. Army. Dual use, both for military and civilian protection, is a clear advantage of these nanotechnologies, because cleaning up the mess is important after an accident such as an oil or chemical spill, just as it is on a battlefield after a chemical or biological attack.

## Human Repair

Since *nanoscience* works close to the interface between man and machine, some terms occasionally slip from one context to another.

Machines start to be described in human terms and humans to be described in machine terms. An excellent example of this, and one of the most promising fields of nano science, is human repair.

One form of human repair involves fixing "*mechanical*" problems with the human body, such as bone fractures, torn muscles and ligaments, burns, and cuts by helping the body heal itself. This differs from the conventional method for replacing bones using steel or ceramic implants, stapling tissue, and using other invasive tactics.

Human repair can greatly accelerate the process of healing. Consider the common case of a broken bone. In order to heal, it will probably have to be immobilized. This can be done with a plaster cast for many simple arm and leg fractures, but if the fracture is complex it may involve surgery bone nails, or even full-body immobilization.

It usually takes months for a bone to set and even longer for it to heal completely, but human repair can change this. Research done in Sam Stupp's group at Northwestern University addresses this very problem. Stupp's approach consists of injecting the site of a fracture with small molecules that assemble themselves into bonelike fibers that bridge a fracture. The self-assembly of these fibers from the liquid can rake just seconds, and as soon as it is in place, healing can begin. Osteoblasts, cells that aid in the development or repair of bones, adhere to the *nanofibers*, and natural bone forms to bridge the fracture. This approach to human repair can result in dramatically faster recovery

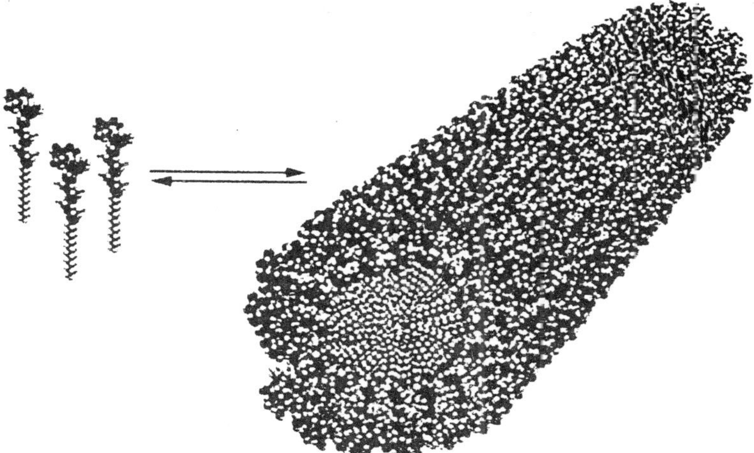

*Fig. 8.1. Peptide amphiphile molecule (left) that self-assemble to form a cylindrical mandrel (right) on which bone can grow.*

from injuries, less surgery, and a better mend when healing is complete. It also does not require steel bolts or implants that must be surgically removed later or that remain with a patient for the rest of his or her life. Almost everyone breaks a bone at some time. Wouldn't it be nice to be able to walk our of a hospital just days after being wheeled in with a broken leg?

Another form of human repair centers on controlling bleeding from traumatic injuries. This innovation will increase our ability to treat mass casualties, and it has obvious dual-use applications.

## Information War

More and more of modern warfare is about who controls information, not just who has the most troops or the best weapons. The information war takes place at many levels, including intelligence efforts before a confrontation, identifying enemy positions, cyber-war fire, and protecting lines of communications. Almost all warfare now has an information component, but we will focus on two or three aspects in which nanotechnology is likely to change the game significantly.

### High-performance Computing

The first and most obvious influence of nanotechnology will be in high—performance computing. It is clear that there is a limit to how much microchips based on current designs can be improved. Already the smallest components are entering the nanoscale, and as they do, engineers are confronting the strange properties of the quantum world. Most projections agree that by the year 2010, Moore's Law, the

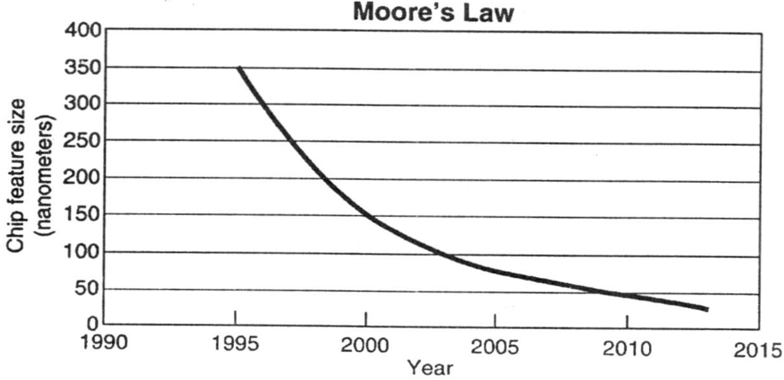


*Fig. 8.2. Moore's first law predicts the rapid exponential growth of the density of transistors on a computer chip.*

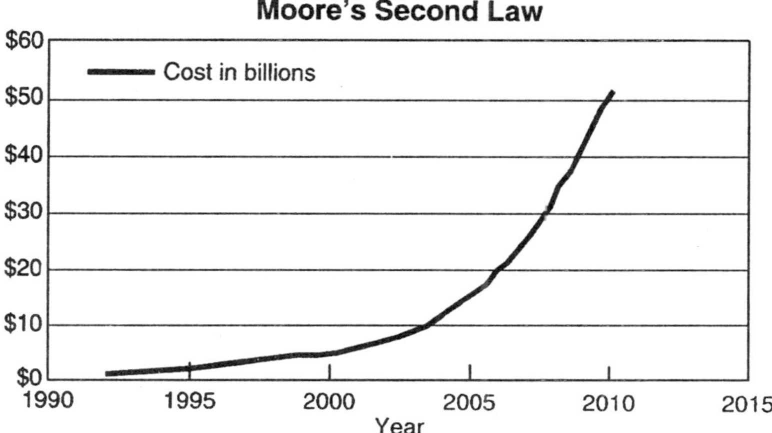


*Fig. 8.3. Moore's second law notes the rapidly increasing price of fabrication facilities for making computer chips.*

venerable computer industry maxim that computer components halve in size and double in performance every 18 months, will be end compound this, Moore's less cited second law that chip fabrication facilities double in cost with each chip generation shows no chance of abating. This means that by 2010 we might not only start seeing dramatically declining improvements in silicon technology but that the fabrication lines to manufacture the most cutting-edge chips will cost between $25 and $50 billion. At that pricey most semiconductor companies would go out of business.

This is one of the chief drivers that has made nanotechnology so hot. Various new approaches to higher performance computing will only be possible using nanotechnology. DNA computing uses DNA to crunch data as it now stores the formulae of life. Molecular electronics uses individual molecules as electronic component. Bioelectronic computing uses proteins and other biostructures as computer components. All-optical computing uses light instead of electrons for all functions of a computer. Perhaps most powerful of all, the quantum computer can compute problems using the quantum properties of matter. These advances have caused nanotechnology research pioneer Stan Williams of Hewlett Packard to suggest that, in spite of forthcoming limitations on current technology, "It should be possible to compute one billion times more efficiently than is currently possible. That means you could hold the power of all earth's present computers in the palm of your hand."

Faster computers are not just essential to the economy. They also allow for a number of security applications. For example, you often show your photo ID to enter a building. Devices exist that can instead scan your hand, eye, or face to determine your identity, check you against a list of restricted people and provide building security by letting you in or locking you out. These devices are much more secure than a quick inspection of a photo ID, but they have a problem: comparing data from a user to a small library is fairly easy, but if the library of users includes tens of thousands or even millions of people even the fastest computers can be swamped. The problem is even more severe if a DNA database of known terrorists is to be developed and used to help track them down. Although this data base effort is controversial, DNA for forensic use has been common for some years.

Image recognition in *satellite photography* is another potential nano application, especially since nanotechnology based computers are showing a particular penchant for pattern recognition. Data storage is yet another—DNA-based storage is one of the most efficient methods for long-term storage where access speed is not an issue, and three-dimensional "*holographic*" (light-based) storage offers quick retrieval as well as storage capacities much greater than anything we have today. Both of these data storage techniques exist already, though they are not yet commercially available. Nanotechnology has already made gigabyte hard drives reality, and that is proving to be a tiny fraction of what the tiny can do.

The loftiest goal of high-performance computing remains *artificial intelligence* (AI). AI would allow robots to take on many of the most risky jobs, such as serving in front-line ground forces. True, self-aware AI is probably still quite far in the future, but machines that can think their way out of complex situations may be just around the corner.

**Cryptography**

Code-breaking is essential to any serious intelligence gathering. If a nation wants to intercept terrorist communications or enemy transmissions in the battlefield, it must have good code-breaking ability Arguably the tide of World War II turned for the Allies when Alan Turing and his colleagues broke the Nazi cipher called *Enigma*. By 1942 the allies were decoding tens of thousands of Nazi communications each month, and the intelligence they accumulated has been estimated to have shortened the war by at least two years (though it is hard to know what effect the atomic bomb would have had). However, Turing's efforts took several years, an amount of time which is impractical in modern warfare and intelligence gathering. Today, advanced *cryptography* is widely available. Keys and encryption schemes can be changed quickly. Terrorists and hostile military forces take advantage of these technologies to make it difficult to recover information from captured computers or intercepted transmissions.

Many modern digital encryption schemes (such as RSA, a common scheme for e-commerce applications) rely for their security on the idea that large numbers are very difficult to factor and that the time it takes to factor them goes up exponentially with the size of the number. While a computer can in principle break these codes using a guess and check technique, breaking today's best commercially available codes using this approach would take a modern PC (circa 2004) more time than has passed since the fall of Rome. For this reason, the common commercial encryption schemes that permit data security within our society can also allow terrorists and other criminals to exchange data with almost perfect security.

Quantum computing is one solution that could turn the mathematics of code-breaking on its head. A quantum computer can, effectively, work on all the possible solutions to the problem at the same time rather than in sequence. In reality the solution is a bit more complicated than that, but there are reasons to believe that a single quantum computer could solve an encryption problem far beyond the reach of the mightiest PC—and it could do so in just a fraction of a second.

Breaking codes is more of a double-edged sword than most military or defense technology. The overwhelming majority of people using cryptography are employing it for lawful and decent reasons including trading stocks, buying goods, or protecting the privacy of their e-mail over public networks. Even more than most nanotechnology, decryption technology with this much power must be used sparingly and its abuse must be punished diligently. Quantum computing has applications far beyond decryption. Searching massive amounts of data and even modeling other quantum mechanical systems are two obvious examples.

**Pervasive Computing**

Relatively little thought has gone into the form, shape, or mechanical properties of modern computers. One or more silicon chips form the core of almost all electronics from computers to music players, and the silicon chips are still centimeter-scale pieces of silicon in plastic packages often trimmed with metal heat sinks. The chips sit on circuit boards of rigid plastic. This is a perfectly good way to build electronics for many applications, but it prevents them from being integrated into clothing, packaging materials, wall paint, and many other places where they could be useful.

It may not be obvious why it is desirable to put computers into wall paint. Many people justifiably feel that computers are already too much a part of their lives, but the applications for tiny, flexible computers (so-called "*pervasive computing*") are very intriguing. For example,. pervasive computing could have household applications, such as controlling lighting, temperature, or music volume based on inputs ranging from what you are doing to the expression on your face.

In addition, pervasive computers have security applications when coupled with appropriate sensors, since they cannot be disabled easily the way motion detectors can. Unlike most security systems, which operate best after business hours, they can so allow the continuous monitoring of an area even during periods of normal activity. Finally they are necessary for many of the soldier applications we've discussed, including the next uniform.

So far the most promising avenue for this kind of capability is a nanotechnology called *molecular electronics*. Molecular electronics involves using individual molecules as components in electronic circuits. This is the ultimate level of miniaturization for electronics of the type that we are used to transistors, capacitors and the like. In addition to being very small, molecules (as opposed to metals or crystalline materials like silicon) are generally fairly soft and flexible. Molecular components may not need to be affixed to a rigid plastic board the way current electronics do. Instead, it may be possible to put them on softer polymer surfaces or other flexible backplanes. This would allow them to he used for many kinds of computing applications.

Molecular electronics will initially be soft and cheap. For example molecular thin film transistors (and we mean *thin*—just a few nanometers in thickness) now being developed by Phillips, Lucent, and others, are designed primarily for disposable applications like identification tags to label soup cans, pieces of mail, or even individual pills. The soup and pill labels would help track possible contaminations, including terrorist modifications, and allow fast recall. Recall the concerns with meat product contamination and cyanide in Tylenol bottles.

**System Diversity and Survivability**

One of the greatest threats to American security is the vulnerability of our computing and communications infrastructure. This comes in several forms. The fact that a single computer architecture (the PC) and a single operating system family (Windows) now comprise the overwhelming majority of our data systems represents a huge threat, since any weak ness discovered in either design may be exploited on millions of other systems. This problem is similar to dependence on a single crop for food. For example, in the Irish potato famine a blight wiped out the whole crop of potatoes and there was no alternate crop, such as corn or wheat, to see the population through the crisis. High-impact computer viruses have shown that the analogy holds for computer systems. What often starts as a prank can result in billions of dollars in damage and lost productivity. While nanotechnology is not currently poised to do anything about operating systems directly, it may help to evolve the underlying architecture of systems, diversify them, and make them more secure.

As mentioned earlier, nanotechnology is already promising at least five possible major revisions of traditional computer architecture. These include all-optical computing, bioelectronic computing, DNA computing, quantum computing, and molecular electronics. Some of these, like all-optical computing and molecular electronics, are general purpose and may work quite similarly to current machines. Others, like DNA and quantum computing, will be much more effective for speeding certain kinds of computations (such as code breaking) and can offer some types of tamper-free computing, but may never find application in general-purpose computing. All of them change some basic rules.

"While we have talked a lot about chemical and biological weapons, we have not spent much time discussing the third major category of weapons of mass destruction—nuclear weapons. There are many reasons for this. Not even nanotechnology will protect a city from a megaton nuclear device. It may be helpful in detecting the weapon or even shooting down a missile carrying one, but there are no technologies on he near-term horizon which will do much to protect us if the worst happens.

Having said that, a nuclear weapon from a less-developed state is unlikely to be an advanced, hydrogen bomb with a megaton yield. Smaller, so-called. "*tactical*" or "*dirty*" nuclear weapons are much easier to make, buy, or smuggle into a tar get zone. They cause a great deal of immediate damage, but there are other environmental effects of a nuclear detonation, which must be dealt with similarly to remediation of chemical and biological weapons. Another effect of such weapons is the nuclear *electromagnetic pulse* (EMP). An EMP is a wave of massive energy that can disable or destroy electronics well beyond the reach of the actual nuclear blast. The military specification for electronics requires ruggedness well beyond most civilian specifications. Not only must military electronics operate in hotter, colder, and damper climates than civilian electronics, they must also have some resistance to EMPs. While it is possible to improve the resistance of electronics to EMPs, it is very difficult to shield them completely, and most shielding requires significant sacrifices in terms of weight and performance. There is reason to believe that all-optical computing and DNA computing as well as some other nanotechnology-based computing options will be naturally resistant to EMPs. They will also have better performance in other hostile conditions including outer space.

All-optical computing (using light rather than electricity to process data) is one example that will not only resist EMPs, but will also be less sensitive to electronic eavesdropping. Since the optical signal is confined to a fiber much more tightly than an electronic signal is truly confined to a wire, it is harder for an external piece of equipment to listen in. The limiting factor to creating an all-optical computer is the design of a small, efficient optical switch to perform the same duty as a transistor. Bulkier optical switches for all-optical communications net works exist, but optimization to the point of usefulness for a complete computing platform has not yet been done. It will be done eventually, and all of the most promising designs to date are based in nanotechnology.

All of the nanotechnologies that we have just surveyed for homeland security exist in some form. Some are already commercial products and some are still several years from development. Even the far-out notions such as quantum computing are close enough to hand that corporations, venture capitalists, and universities are becoming interested. For example, D-Wave, a Canadian startup interested in practical quantum computing, raised a large round of private equity under the most difficult conditions at the beginning of 2003.

There are dozens of other ways in which nanotechnology could affect homeland security, and many of them will only be recognized as fundamental research continues. Unlike defense technology, however, homeland security technologies will impact all of our lives directly. Some, such as human repair, will only change life for the good. Others, such as advanced cryptographic technology, could be much more threatening. And still others, like AI or pervasive computing, could change our day-to-day lives as much in the next 15 years as electronics, the Internet, and telecommunications have done in the last 50.

# 9

# MEDIA AND SOCIETY

*Improving society* is a difficult task. More generally, improving complex systems is a difficult task. If you cannot figure out which way is up, see if you can figure out which way is down. Doug Engelbart, back in the early 1960s, wanted to explain to people why interactive systems would make a significant difference to their lives, and to their ability to express ideas. The origin on the axis is what people were doing at the time—writing with pencil and paper. When he found himself unable to communicate to people how much better things could be, he contrasted their current experiences with how much worse things could be. He tied a pencil to a brick, handed it to people and said, "*Okay, now write.*" People found it very difficult. The unwieldy nature of the tool interfered with their ability to express ideas. With the pencil and brick for contrast, he effectively asked two questions: "What made the difference?" and, "How can we move further in the other direction?" This experiment showed people how important their tools and their media were to their effectiveness, and helped them start to see the next brick to remove.

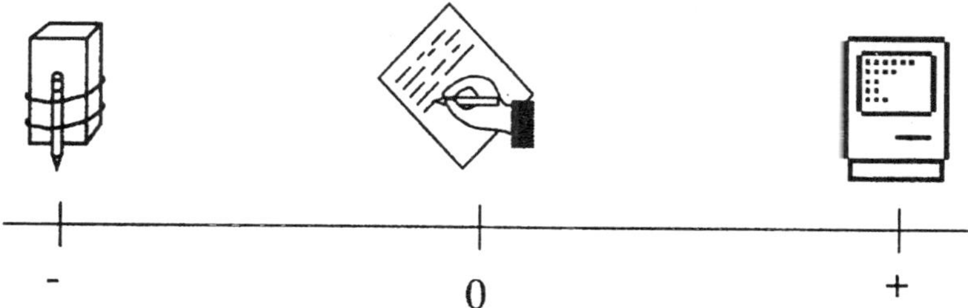


*Fig. 9.1. Engelbart's pencil and brick experiment.*

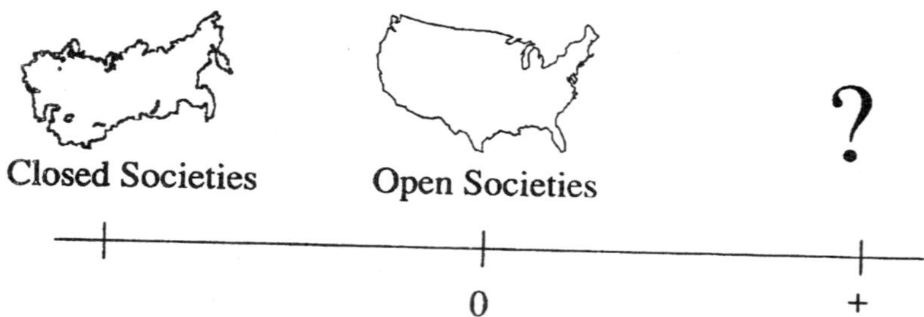


*Fig. 9.2. Marx's pencil and brick experiment.*

Karl Marx performed a similar experiment on society over the course of most of this century. The origin represents where we are now. Karl Marx tied a very large brick to a very large pencil and the last few years have revealed the result to be far worse than the even his harshest critics imagined. What made the difference between the societies? Two important elements were open markets and open media. How can we move farther in the other direction? In this presentation, I will be addressing the nature of open media, how they differ from closed media, and how social hypertext systems can enhance the advantages of those media. Applying information technologies to the further opening of markets is left as a mission for the reader.

## Media Matter

*Media matter*, because it is in media that the knowledge of society evolves. The health of the process by which that knowledge evolves is critical to the way society changes. Karl Popper, the *epistemologist*, had the insight that knowledge evolves by a process of variation, reapplication, and selection, much as biology does. "*Variation of knowledge*" is what we call "*conjecture*"—hypothesis formation, tossing new ideas out there. "*Replication of knowledge*" is the spread of ideas through publication and conversation. "*Selection of knowledge*" is the discrediting of conjectures through the process of criticism. The ability of our knowledge to progress over time depends on an ongoing process of criticism, and criticism of criticism. The ideas that survive the critical process tend, in general, to be better than those that do not.

In closed societies, when arguments cannot be spoken, hard truths cannot be figured out. When people cannot openly criticize, cannot openly defend against criticism, or cannot openly propose ideas that conflict with the official truths, then they are left with mistrust and cynicism as their only defense. This leads to the simple heuristic of assuming the official truth is always wrong. For example, because science was promoted by the Soviet propaganda machine, pseudo-science is on the rise in Russia. Because anti-Nazism was promoted by the East German propaganda machine, Neo-Nazism is on the rise in East Germany. The official truth is neither always right nor always wrong. Society needs a more sophisticated process for judging claims.

Our society does have open media. Are we in the best of all possible worlds? Are our media good enough? Can they be made significantly better? Among our media, TV is so bad that it is a joke. Only slogan-sized ideas can be expressed. We prize the quality of discourse in our books and journals, but critical discussions in them are only loosely connected. Starting from the expression of an idea, it is hard to find articles that criticize that idea. When arguments can not be found and navigated, the next harder truths still cannot be figured out.

## Links

*Hypertext links* are directly inspired by literary practice. Literature has many different kinds of links connecting documents into a vast web. Textual examples of these links include bibliographical references, marginal notes, quotation, foot notes, and Post-it notes. We propose to build engines of citation, so that people can navigate this vast web of literature at the click of a mouse. Most computer text systems are predicated on a misconception that the meaning of a document is represented purely or primarily by its content. Documents are not islands. Conventional computer text systems put their effort into the appearance of individual documents. A context helps answer questions such as, "What were the ongoing controversies that the author had in mind?" "What views was he supporting or attacking?" "What attacks was he guarding against?" We must understand this whole web of connections in order to understand the documents we are reading. The Xanadu system is built to provide as much support for this contextual information as for content. With the ability to follow the links in this vast web of documents, is it not easy to get lost? How does one stay oriented? One answer to these questions is *guides*, a new kind of document that provides an orienting view together with links into the existing literature.

## Hyperlinks

Because "*nanotechnology*" is now used by many to mean any technology approaching the nanometer scale, we have been forced to retreat to the term "*molecular nanotechnology.*" Hypertext terminology has gone through a drift similar to nanotech terminology. The Xanadu project is the one that coined the term "*hypertext*" and originated the notion of the hypertext "*link.*" However, because the term *link* has come to be viewed as something much less capable than what we meant by it, we are now calling it the *hyperlink*. The distinction between the link and the hyperlink is crucial for supporting active criticism in open media.

*Hyperlinks* are fine-grained, bidirectional, and extrinsic. Frequently, an argument is not with a document or chapter as a whole. It is with a particular point that someone made at a particular place in the text. For example, someone refers to the fourth law of thermodynamics, and someone else writes a criticism saying there is no fourth law of thermodynamics, linking it to the original. The fine-grained property allows the link to designate the particular piece of text with which one is taking issue. Bidirectionality enables readers of the original document to find the criticism, enabling them to exercise fine grained skepticism, and to constantly ask themselves, "What is the best argument against the thing we are reading *right now*?" and then, "What is the best argument against that, in turn?" Links provided by other hypertext systems generally have been only in the forward direction, enabling a reader to find those documents *referenced* by a given document. However, to find criticism, the reader must find the documents that *refer to* the documents they are reading.

*Extrinsic linking* is the ability to link into a document without editing it. Several other systems support the creation of links that are fine-grained at the targeted end, but these others do so only by modifying

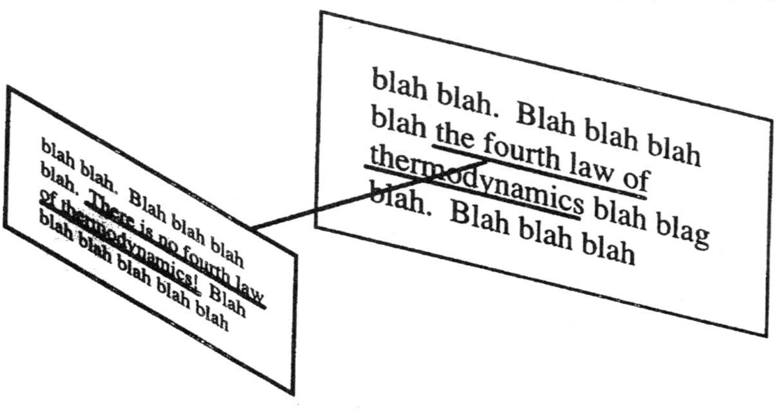


*Fig. 9.3. Hyperlinks.*

both source and target documents. Critics normally will not have the ability to modify the documents they are criticizing. They could spin off their own version into which they attach these links, but then other readers still cannot find these criticisms from the original documents.

Part of what we mean by "*open media*" is that everyone who is connected to the system can read what they are permitted to read, can write new things, and can make them accessible for others to read. This includes making links to any thing that they have read, so that anyone else who reads the original can find the material that has been linked to it. All readers of the system are potential authors. We can think of this process as active reading. Frequently, people make marginal notes to themselves. This is a medium in which readers can share such things with each other. When much writing is commentary about other text, the commented-on text is the best rendezvous point for the authors and readers of commentary to find each other.

## Emergent Properties

This kind of accessible criticism can provide for decentralized consumer reports. When people post on the system documents that are either products or descriptions of products, customers of those products can post criticisms of them. What did they think of using them? This commentary can guide the purchasing decisions of others.

A particular capability we are used to in conversation is hearing the absence of a good response to an argument. A reader not only can see what the most compelling arguments are against some statement, but also see when there are none, or when all the seemingly compelling arguments have been successfully refuted. Such absences are quite obvious in conversation. Electronic media can make these absences obvious as well, but in a context where the absence will be much more telling, because the missing argument could have come from a much larger audience over a more extended period of time.

Other hypertext systems with their unidirectional links reproduce the asymmetry present in our paper-based media—it is much easier to find something that a document cites, than it is to find those documents that cite a given document. One of the effects of this asymmetry in paper media is the pathological division of scholarly fields into disjoint "*schools*." Instead of healthy intellectual engagement, debate, and cross-fertilization of ideas, we see a process of increasing inability to communicate between schools, and more preaching to the converted within a school. The terrible irony of attempting scholarship with unidirectional links is that the *very attempt to engage in healthy debate across schools accelerates the pathological division process*. How does this occur?

Let us consider two schools within a discipline. Generally, students within a school see the documents supporting the positions of that school. The students also see criticisms of documents in the other school. Intellectually eager and honest students, seeking to know both sides, occasionally will follow these criticism links forward. The result is that they will see the parts of the other school's literature that is *most soundly criticized* by their own school, immunizing them more and more against the foreign ideas. With bidirectional links, these students can also find the greatest challenges to their own school. Bidirectional links also enable them to find the *most telling criticisms* of the ideas they are inclined to accept.

### Transclusion

Before there were modem economies, there were many little villages, each with their own little manufacturers having to go through a large amount of the production process themselves. These economies were, therefore, much less productive. An individual baker or shoemaker, for example, would reproduce the same kind of work that was being reproduced in many other villages, and would have to fashion a shoe, not quite from raw materials, but without intermediate goods. In extended economies, people can build on one another's work, and there can be a finer-grained division of labour and knowledge, with better specialization.

Now, with respect to literature, authors are frequently faced with the task of re-explaining and restating background material that has been explained well elsewhere. If you could just borrow that material, those existing good explanations, and incorporate them (with automatic credit where due), your efforts could be spent stating what is new. We introduce the concept of *transclusion* to separate the arrangement of a document from its content. There is an underlying shared pool of contents, and all documents are just arrangements of pieces from that pool. The three-circled appearances of the same text are actually just one piece of text in the underlying shared pool of contents, and it just happens to appear in three different arrangements, which constitute three different documents. We refer to the three documents as *transcluding* that piece of text. The separation of content and arrangement also leads to good support for incremental editing. Different versions of a document are just different arrangements of mostly shared content.

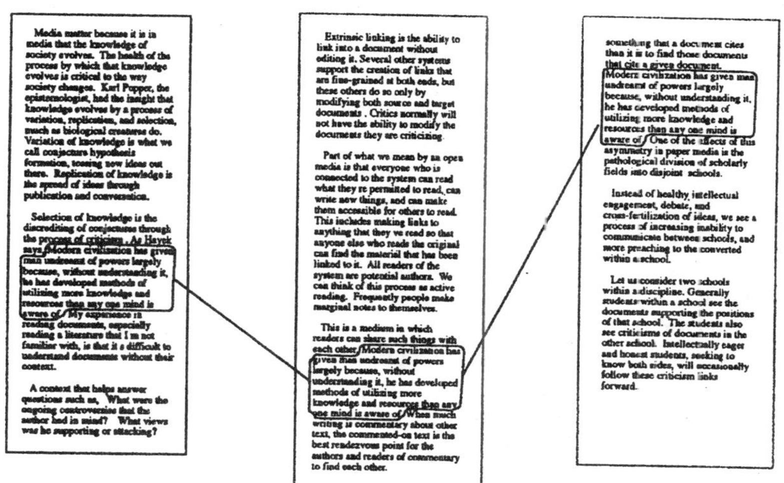

*Fig. 9.4. Transclusion—different arrangements of shared contents.*

This is more than just a hack to avoid the storage cost of making separate copies. *Hyperlinks* are linked to the content, not to a span in an arrangement. Therefore, when someone writes a criticism of content as it appears in one arrangement, that criticism is visible for the same content as it appears in all other arrangements, including arrangements that were made before the criticism was attached. The normal incremental editing process of a single document is analogous to *evolution by point mutation*. The ability to translate text from other documents allows the analog of sexual recombination. Were links visible only from the arrangement into which they were made, both variation processes would destroy selection pressures by leaving criticisms behind.

## Historical Trails

As you are editing, an *historical trail* gets left behind—bread crumbs in history space. The historical trail is simply a sequential arrangement of the successive arrangements of contents. This is yet another kind of context important for understanding. "Things are the way they are because they got that way." Understanding how they got that way often aids our understanding of what they are.

## Detectors

In addition to looking into the past, one also reads a literature knowing it will be changing. How can one keep up? To keep track of what is happening, to keep up with changes, we introduce *detectors*. One can post a *revision detector* to find out when things are edited, when new versions of something appear, and then one can use *version compare* to find out how they are different. With version compare, one can engage in *differential reading*—reading just the differences between the current version and the version most recently read.

*Link detectors* are a way of finding out when new links are made to existing material. Let us say that you published something, and you want to find out when others post comments on it. You would like to be informed of comments, but you do not want to have to go back and constantly recheck all the things that you have written, so you post a link detector on all the things that you have writ ten, as well as on other documents on which you are interested in seeing further comments. You want to see what people will say about them. As new comments are posted on those documents, you are continually informed.

E-mail is just the special case where you establish a canonical point in the literature, for each person—a place others link to in order to send that person a message. That person simply has a link detector there saying, "Show me all new things that are attached to here." This generalizes to treating any shared point of interest in the literature, as in some sense, a "*mailbox*," or a "*meeting room*" for further conversation or conferencing about a topic. Canonical documents become meeting places. Should two disjoint discussions about the same topic spontaneously form in two places, anyone who notices can just make a link between them. The link detectors of each community will then inform them of the existence of the other.

## WidgetPerfect Saga

This is a true story about how a *hypertext system* was able to save several thousand jobs. One special characteristic about this true story is that it is a true story from the year 1997. It is a story about one of the events that took place at the company—most of you have heard of it—called *WidgetPerfect*. *WidgetPerfect* is the second largest manufacturer of widgets in the world, second only to their big competitor, Microwidget. The people at WidgetPerfect in the year 1997 had identified a really significant opportunity in the upcoming expanding environment of widget components technology.

They were developing the world's first fully modular widget. They had a team working on it. Dan was in charge of the preparation of the marketing materials for the modular widget. Ruth was in charge of the technical work team, and John was in charge of the budget and finance, as well as all the costing. At this point, the modular widget was in prototype stage when a very unfortunate thing happened.

Microwidget, the big competitor, came out with a partially modular widget, hitting the marketplace first with an inferior product. It was technically inferior, but nonetheless it was in the marketplace first.

Dan was examining this *Microwidget*, partially modular widget, and it was overall inferior. Nonetheless, it had one really striking improved feature. It had a funculator made out of titanalum, whereas the fully modular widget that was being developed by Ruth only had a

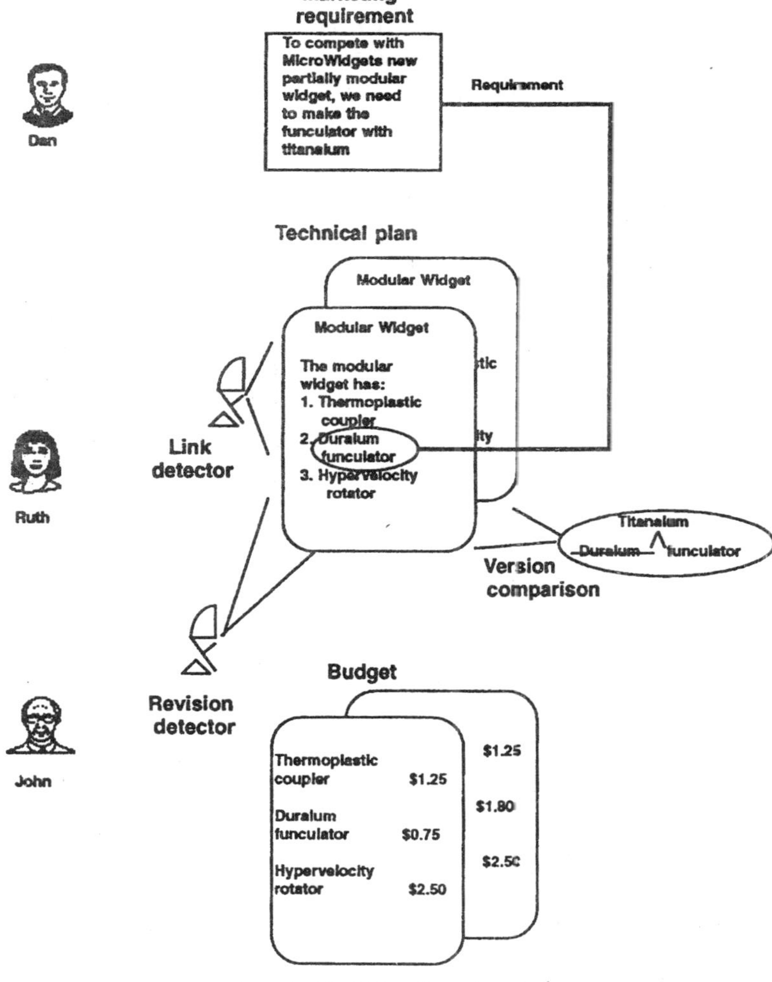


*Fig. 9.5. The WidgetPerfect scenario.*

duralum funculator. This was an important improvement for certain key market sectors. Even though the partially modular widget did not have anything comparable to a thermoplastic coupler or a hypervelocity rotator, they had to make a change. So, Dan created a new document in the marketing requirements describing this titanalum funculator. He attached a link to the part of the technical plan that specifically referred to the duralum funculator in the current plan. He made that a new requirement.

We have good news.

Now, Dan knew that in order to get anything to happen with improving the widget prototype, he would have to talk to Ruth. He was reaching for the telephone to call Ruth when Boeing, the largest purchaser of widgets in the world, called him about a $15 million widget order. He got distracted with this purchase, and he never quite got around to calling Ruth.

Ruth, knowing that the success of her technical design depended on her being able to respond promptly to new requirements, had attached a link detector to her technical plan. This link detector would be constantly watching for new links of the link-type *requirement* to be attached. When Dan attached the new requirement to the duralum funculator, Ruth's link detector went off. Ruth was alerted. She followed the link detector out to the link, followed the link back to the new requirement, saw what the required change was, and modified the technical plan to reflect the use of a titanalum funculator.

Well, this was all very fine, except for an additional problem. As I think everyone here knows, titanalum is considerably more expensive than duralum, and so this had some significant effect on the manufacturing cost. Ruth knew that this would have an impact on the budget, and she was reaching for the telephone to call John when smoke started billowing from the laboratory where the prototype of the modular widget was being manufactured. She ran off to deal with the emergency and never quite got around to calling John.

We have good news.

John, knowing the success of his budget was completely dependent on his responding to modifications to the technical plan, had attached a revision detector to the technical plan and this detector was constantly watching for updates. So, when the technical plan was indeed updated, John's revision detector went off. He followed the revision detector up to the technical plan, used the hypertextual version compare capabilities based on the transclusion relations, compared the new version of the plan to the old, and found that the change was deleting duralum and replacing it with titanalum. He then went back into the budget and updated the budget documents to reflect the increased costs caused by the use of titanalum.

As a consequence of this, the modular widget program was completed on time with a fully adequate specification. It was a completely superior product. It blew Microwidget off the face of the Earth. As a consequence, thousands of jobs at Widget Perfect were saved.

## PERMISSIONS

A social system, to a large extent, is a system of rights and responsibilities. Xanadu has an extensive permission system called the *club* system, intended to deal with some of these issues. Bob has sent it as a mail message to various people in a blind carbon copy ("bcc") relationship. Alice and Chuck are both members of the *bcc* club of people who have permission to read this document. Bob, though, is the

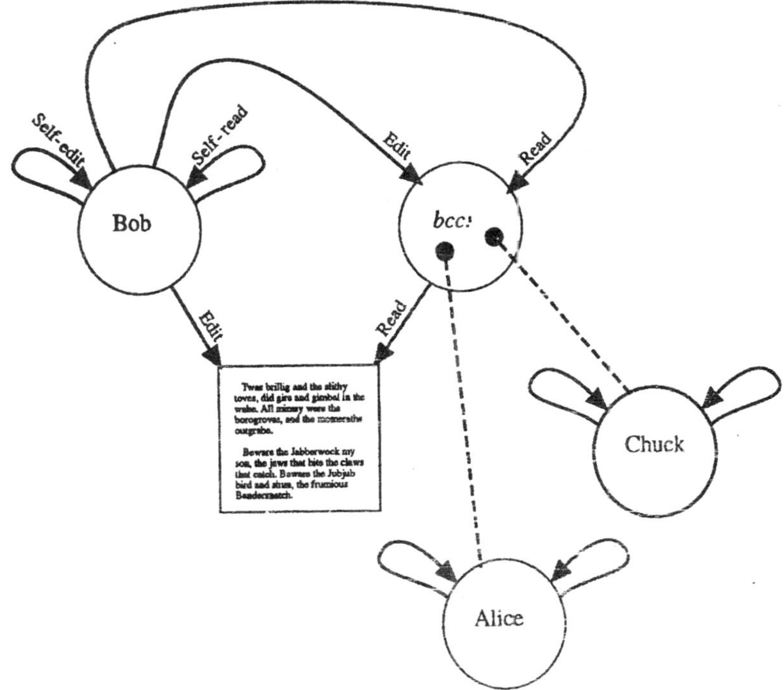


*Fig. 9.6. The club system.*

only member who can read or edit the *bcc* club. If this were a *cc* list, Bob would still be the only person who could edit it, but it would be self-reading. Everybody who was a member of such a *cc* club could see who else was a member of that same club.

This demonstrates a principled answer to permissions *meta-issues*—one can distinguish between who can read a document, who can read the list of people who can read a document, and who can read that list, out to any desired degree of distinction (and similarly for the editing dimension). However, infinite regress and needless complexity are avoided by using clubs that are self-reading or self-editing (or both) whenever further distinction is currently not necessary. Should such distinction later become necessary, it can always be introduced by someone with appropriate edit permission to the club in question. Users only grow meta-levels on an as-needed basis.

Our permission system also supports the notion of *accountability*. All actions in the system are taken by *someone*. When you look at information in the system, you see some identity attached to the actions taken on the information. There are no official truths. There is only who said what, and the structure of the system reflects that.

### Reputation-based Filtering

One of the potential pitfalls of an open hypertext system is the junk problem. The ability to find good commentary and criticism will be especially important when reading very important documents, but it is precisely on these documents that one expects to be inundated with tons of worthless or irrelevant links. Without a filtering mechanism, it would be on exactly the documents for which one most needs good commentary that the provision of commentary would be most useless. For example, imagine how many links there would be onto the First Amendment to the Constitution.

Links can be *endorsed* as worth reading by various readers. However, no one may endorse with the identity of another. Different endorsers will establish varying reputations with different readers, much as with movie reviewers. Readers can filter their views of links into a document both by who endorsed as well as by link-type. When even this mechanism gives too coarse an answer, one can rely on documents such as a hypothetical *Guide to the Citations to tile Bill of Rights*, endorsed by a reputable publishing house. This very same link filtering ability is also what allows one to find such guides in the presence of a swamp of links.

## Hypermedia

Increasingly, ideas are being expressed in media other than text, and increasingly, computers are used to handle these other media. We usually refer to *hypertext* because text is the most important case and the clearest example. However, none of the things you have seen the system do, is in any way specific to text, or even to media that have linear flow to them. It all applies equally well to a variety of other media (such as sound, engineering drawings, Postscript images, and compressed video). In all cases, one can make fine-grained links, edits, transclusions, and version compares (even if the data is block-compressed or block-encrypted). Although the implementation has some optimizations targeted at text, in no way does the *architecture* make any special cases for text. Documents can, of course, be composite arrangements in which several media are mixed together.

## External Transclusion

No software system is an island. We do not imagine that once the product is available, everyone will instantly take all information to which they want access and transfer it into Xanadu. We have to coexist with many other systems for many good reasons.

We handle that with *external transclusion*. Our documents are able to transclude into arrangements that are within the system. These, in turn, are able to represent transclusions of materials that are stored elsewhere. By perceiving other systems through the window of Xanadu, you can see those other systems as if all those documents were within the Xanadu system. Through Xanadu, it could follow a link from a WAIS document into a Lexis document, even though neither system has any notion that such a link even exists. It is not just that the Xanadu system is not an island, that we have to coexist with everything else, it is that through Xanadu, those systems are able to coexist with each other in a way they are unable to now, making them into nonoislands.

## Conclusions

When we started building the system, we were thinking purely in terms of paper-based literature—of writing. What we have built is something that has many of the best aspects of both writing and conversation. Many of the aspects of each are complementary. Many virtues of conversation make up for flaws in writing and vice versa. We found ourselves building a system that supports the dynamic give-and-take of conversation, and the persistence and thoughtfulness of literature.

**Table 9.1. Combining the best of writing and conversation**

| *Writing* | *Conversation* |
|---|---|
| Persistent | Visibility of arguments |
| Cumulative | Can "hear an absence" |
| Large expressions | Small expressions |
| Revision before publication | Revision after publication |
| Large audiences | Small audiences |
| Freedom of entry | Fast publication |
| Reputation-based filtering | Feedback from audience |

Our status is that we currently have a working, portable server. It has some bugs in it, including some performance bugs, but we are working on it. We are continuing ahead with the effort on both the server and the front end. The front end is in a preliminary stage. We consider it adequate to show that the server is real, and to exercise its features. We plan to do a much better front end.

The protocol between the front end and the server is very stable, and has been stable for a long time now. Our plans are to get investors, and to finish both the front end and the server. The target for our first product is small- to medium-sized work groups within companies that have a large body of documents they need to be managing and evolving.

Our first product lacks one major feature. We provide hypertext because documents are not islands. We make the system interpersonal because people are not islands. We provide for the transparent windowing into other systems because no product is an island.

However, for the moment, each server is still an island with respect to the other servers, and so each workgroup is also an island. We have designed the system so that, soon after first product, we will be able to weave all the servers together into a transparent distributed system. When you follow a link from one document to another, if the other document is not here but in some server in Tokyo, it will be transparently fetched for you, and the only thing you will notice is that following that link took longer.

For any media to radically improve the process of opinion formation in society, we believe it needs features equivalent to fine-grained, bidirectional, extrinsic, filtered links. These links must not get lost when the documents to which they are attached change. Issues of authority, privacy, and responsibility must be handled in a robust and secure fashion.

Open entry of readers and editors is chemical for open discussion. Open entry of server providers is less obvious, but equally important, in order to make centralized control impossible. We will be providing support for people who want to do online services based on our soft ware. All of this is necessary to achieve our open electronic publishing dream. In so doing, we hope to improve the quality of public debate, in order to obtain the benefits of the open society yet again.

# 10

# A Story of Real Danger

To an extent, this was a legitimate concern. The heart of a nuclear fission device (bomb or reactor) is a quantity of fissionable material such as U235. U235 is an *isotope* of *uranium* that constitutes about 1% of natural uranium (most of the rest being U238, which is not fissionable). So even if you have a chunk of pure uranium, you need to separate the U235 out in order to have something useful for a bomb.

It is much harder to separate one *isotope* of an element from another than it is to separate two different elements or chemical compounds. Typically, different elements will have different chemical or physical properties. So you put your mixture—ore, for example—through some process that has a different effect on one than the other. A simple example is using a still to separate alcohol from water. Alchohol boils at a lower temperature than water, so if you heat the mixture to somewhere between the two temperatures, the alcohol boils off to be collected in your condenser, while the water stay behind.

But with isotopes, the properties are so similar that it is virtually impossible to distinguish atoms of one from the other. Uranium itself is a very heavy, fairly inert metal. Boiling it is out of the question. To separate U235 for the original bombs during World War II, the United States built the Oak Ridge plant, a city-sized processing facility that still took months to produce enough U235 for one bomb.

The bottom line is that isotopic separation is one of the major technical challenges that forms a serious barrier to entry into the nuclear weapons club. And that's a good thing. Now it turns out that if you only have a few atoms of uranium, or anything else, it is not all that hard to separate them. The difference between U235 and U238 is about 1% in weight. Ionize your atoms and run them through a mass spectrometer, which is an instrument that fires them through a magnetic field. The field makes them curve. The lighter ones curve more easily, the heavy ones don't curve as much; so they arrive at different spots on the other side of the field and are thus separated. The mass spectrometer is a scientific instrument that is very useful in analyzing tiny samples of a substance, but it is of no use for the kind of quantities you'd need for a bomb.

With *nanotechnology*, you don't even need to fool with the tricks like ionization and so forth that bulk-technology mass spectrometers use. Essentially you just build a gadget that grabs an atom and weighs it. If its atomic mass is 235, it goes in bin A, and if it is 238, bin B. Gravity is of no use for weighing atoms, but a centrifuge would work fine. Or more precisely, you build a million billion such gadgets, together with the machinery to distribute atoms to them and put the results back together. This composes a machine that sucks in 5,000 pounds of chemically pure uranium dust on one end, dumps 4,950 pounds of depleted uranium dust out the bottom, and produces a 50-pound sphere of U235 at the business end. Woops, better make that several pieces that could assemble to make a 50-pound sphere, and make sure you keep them apart.

## Self-replicating Weapons

There is a notion that nanotech weapons will be bacteria-like replicators that can live off the land of the enemy's territory and ultimately reduce it to dust or have any other devastating effect that the weapon-wielder desires. There are two points that need to he stressed about such a scenario: First, the weapon scenario is much more likely than that of some kind of accidental replicator release. The difference is that commercial nanomachines will be made for high efficiency (and profitability), and thus be engineered to run fast, from high-energy foodstocks, be controllable, and produce useful outputs. Trying to live off the natural world violates all these desiderata. Accidentally building a nanomachine that would operate in the wild would be about as likely as accidentally building a car that sawed down trees, cut and split them into firewood, and stoked a boiler firebox with them.

It would be possible to design such machines on purpose, however, and weapons have always been the application where efficiency and other such considerations are last on the list. If someone designs a Nanomachine capable of operating in the wild, it will almost certainly have been done as a weapon. The second point, however, is that while it would have possible to design army tanks that burned wood and thus could a in the forest, no one ever did. In fact, actual weapons system strongly in the opposite direction. The reason is that although efficiency is at the bottom of the list of design criteria for war machine power and speed are at the top, and stopping to eat grass or whatever is not the way to get them. Commercial nanomachines will mature more of themselves from high-purity; high-energy raw material and so will military ones.

There just is not the energy available for the taking in the natural world to power machines at the level our current-day machines operate, much less nanomachines. Wood-burning machines cannot even compete with coal-burning ones, much less oil- or hydrogen-burning ones. Although it is too far out and too technically uncertain to mention in other connections, it would not be surprised to see nuclear-powered military machines (radioisotope, not fission). Note that the military currently uses (fission) nuclear-powered ships, which are not cost-effective for commercial use. Where today we have intelligent bombs and missiles, nanotechnology could give us smart bullets. Adding AI in nanocomputers to the mechanical capabilities would give us robot soldiers that could fold up to the size of a rat to creep through cover unfold to the size of a human to manipulate objects or weapons and fly at: hundreds of miles an hour. They could be virtually invisible using phased-array optics skins. Their motions could be too fast for a human eye to track. To be prepared for war, a nation need only have a widespread nanomanu-facturing base that could produce such robots, with matching vehicles and weapons, should the need arise. When push came to shove, the actual army could be produced in under an hour.

A good-sized patch of phased-array optics could focus light on a target. If the roofs of buildings were covered with these, they could act as solar collectors, radars and optical cameras, spotlights, and if tilted enough, streetlights. If necessary, and if used in a coordinated way from many buildings at once, they would function as laser like antiaircraft artillery, In a post-9/11 world it is regrettably understand able why that might be desired. Private spacecraft, in particular, would be viewed with concern, and might be allowed only if there were some "instant-on" distributed air defense.

## WAR

It the twentieth century, the cause of human disaster that clearly stood out above all the others was war. Technology clearly contributed to the deadliness. Will nanotechnology make war worse, or more likely? After all, a nano weapon can't kill you any deader than a hydrogen bomb. What's more, since the development of nuclear tipped ICBMs, a remarkable thing has happened. Nations have quit going to war against other nations that are armed with them. The author is of the opinion that this has a lot to do with the act that such weapons put the political leaders of the attacking country in direct physical danger. As Ambrose Bierce quipped, the meaning of *rear* in American military matters is that part of the army closest to Congress.' Nanotech weapons will only exacerbate this phenomenon making it easier to target the politicians. If so, war could suffer from a further decline in popularity.

Another phenomenon to consider is the industrial revolution and the *Pax Britannica*. If one nation or aligned group does attain a level of technological dominance comparable to Britain in the nineteenth century a similarly peaceful period might ensue. At the moment, the United States and Europe account for a large majority of known nanotech efforts. Although certainly not perfect, the western democracies would probably make a reasonably enlightened and stable world hegemony. The one scenario that seems to merit some concern is the opposite—where other countries develop nanotech first. The United States might then engage in some serious saber-rattling rather than slip quietly behind. Here is a vignette that, thankfully, is completely imaginary at the movement:

It's 2015. In the United States, business as usual has been allowed to prevail. Interest in science has continued to decline. Virtually all the scientists and engineers our universities produce come from, and most return to, other countries. Funding for research is mostly for medical applications, and that is mired in political debates over stem cells and choked with red tape attempting to make it totally safe.

Meanwhile, China has pushed ahead on a broad range of fronts and has produced Stage III replicators. Products begin to appear from China that cannot be made economically anywhere else. No official notice is taken in the United States because our labs can still produce better stuff in expensive, one-off; form. The Chinese are accused of "dumping" and some nanotech products are banned.

China proceeds to Stage IV and Western technology begins to look distinctly second-rate. They are rumored to have engineering design supercomputers. The latest generation of Chinese jets and spacecraft has significantly better capabilities than ours and was designed and produced in half the time. The US military sounds an alarm.

The administration undertakes a crash program to demonize nanotech as "weapons of mass destruction" and get a UN resolution prohibiting it anywhere in the world. This stalls in debate and goes nowhere. The United States makes a unilateral ultimatum to China demanding a halt to all nanotechnology China demurs, and announces that if attacked with nuclear weapons, it will release aerovores into the atmosphere. It is perhaps worth noting that during the industrial revolution, France engaged in a series of wars—that were ultimately won by England at Waterloo.

## Terrorism

On September 11, 2001, in a limousine on the way to Newark airport was observed. Many persons had seen, watching television at home, a plane flown into the World Trade Center. The World Trade Center looked like a pair of smokestacks. The question of terrorism can be divided into two parts: why do people try it, and how do they succeed? The first is a constant of human nature: take people and make them feel powerless, put upon, exploited, and having nothing to lose. Religion is often involved but any strong ideology will do. Some of them will explode in whatever way they can. It is the nature of the beast. The second is more complicated, depending on technology and other circumstances.

If instead of concentrated, regulated, commercial air travel, all the passengers flying that day had been in private air cars, September 11 would have amounted to a handful of carjackings, if that. Personal air craft the size of cars simply could not have brought the towers down. Al Qaeda did not field enough operatives to destroy the towers on a piecemeal basis. You may remember they had already tried explosives-filled vans some years before.

That is not to say that private aircars would have been economically feasible in 2001, but to point out that decentralization is one of the best defenses against terrorism. A terrorist is at his most advantageous position when the defense has made a whole population helpless in hopes of disarming whatever terrorists may be among them. Public transportation is a favourite terrorist target: airliners in the United States, subways in Tokyo, buses in Jerusalem, trains in Madrid.

Oppression, or perceived oppression, generates the kind of anger, shared by a whole community, sufficient to marshal significant resources and motivate suicide agents. In the short run, this is purely a matter for statesmanship. In the long run, the kind of independence nanotechnology can provide may help. In the medium term, the problem of making the transition is delicate and difficult, but must be faced; because in the long term, some terrorists will get nanotech weapons anyway. If nanotechnology is a widely based main stream development, the effects are not likely to be worse than what the people have seen. If nanotech is decentralizing capabilities are used, terrorism could get harder to pull off. It would also be easier for the mainstream to retaliate. As of this writing, after more than two years and after conquering two countries on the other side of the world, American forces still have not captured Osama bin Laden, despite a $50 million reward. Some of the more useful drones: *semirobotic aircraft*. With nanotechnology we could hood the country with drones the size of flies. Hiding would be much more in ore difficult.

Current-day drones have been equipped with air-to-surface missiles and used to strike buildings and vehicles. *Nanodrones* could carry stings of various kinds and target individuals for lethal, debilitating, or punitively painful doses of chemical or biological agents.

A final note on terrorism: In my opinion, the major high-tech terrorism threat is not nanotech, but biotech. Not only have not we caught bin Laden, we have not caught the anthrax mailer, either. He or she seems likely to have been an American working alone. Alone!—while bin Laden and the Taliban had thousands of followers and supporters.

The *nanotech version* of a bacterium, a loose mechanical replicator of microscopic size, might be more efficient than a natural one. But there can be no nanotech version of a virus. A bacterium carries its own building machinery, which nanotech can make in a more efficient form. But the virus is even better—it carries no machinery at all. It uses the machinery in your own cells for its purposes. It consists of, in simple terms, just the genes needed to make copies of itself, and your cells do all the work. That is a level of finesse nanotech could not use. But note that viruses could be made right now with DNA you can order online—$2.35 per base pair—and with equipment in a high school bio lab.

As far as technological timescales are concerned, *biotech* is already here. It is as easy or easier to hide than nanotech. And people have not come close to realizing the scope of threats it could produce. A properly tailored superflu could kill millions, given, say, twenty suicide circle agents to start the spread all across the country at the same time. Or someone could develop a mad cow prion for chickens. Or a Cipro-resistant anthrax. Or transplant the botulin toxin gene into algae and drop samples into reservoirs. Although these are not nanotech threats, nanotechnology (nanomedicine in particular) would be invaluable to help fight them.

## Economic Status

In the foregoing, we have addressed the applications of nanotechnology as if it were the United States and/or Europe that developed and deployed it. However, that is far from a certain outcome. Although we have a great head start at the moment, we could easily wind up as the Scotland of golf, the England of tennis, or the France of the industrial evolution. The reasons, as explained earlier, are manifold: bureaucratic business as usual here; a decline in the number of Americans going into science and technology juxtaposed with a huge increase in the number of science and technology students in Asia; and, probably, a greater willingness to take risks there.

Add to that, of course, the fact that Asia is not monolithic, but has numerous competing nations. This has several effects: competition is a strong motivator and acts as a spur to prevent bureaucracies falling into the lassitude which is their normal state. Look at what NASA achieved in the 1960s, when there was a "space race," compared to what it has done since. Second, in a project of research, one country might commit to one approach, and another, knowing it'd be behind if it just followed suit, could try something else that might turn out to work better. But the advantage is more subtle than simply trying different things.

In a hierarchical organization, when a contrarian scheme begins to look like it might have been a better idea than the reigning orthodoxy, there's a strong temptation on the part of the higher-ups to quash it. In a situation where the competing elements are separate nations, this can be avoided to some extent.

If nanotechnology were developed in Asia with a significant lead time, the United States would be reduced to the status of a second-rate power. We are used to living behind the buffers of the great oceans and projecting our power where we wish on the globe. We could lose both—the dominant power might simply decide to tailor the Earth's climate to suit itself, for example, without bothering to ask us. You can be sure that the economic realities would shift, and we would be on the wrong end of some bad deals, shut out of monopolies, and, in general, have a lower standard of living and fewer opportunities than if we maintained the edge.

## NANOPARTICLES

One danger that has been mentioned in the press in connection with nanotechnology is the release or biological effect of *nanoparticulates*. Today's nanoscale technology is producing some new products, such as paint that resists bacterial growth. But the effects of such substances, while different in detail from current substances we deal with do not differ in any significant overall way. After all, ordinary untreated wood contains substances that inhibit bacterial growth. Ordinary chemicals are magic of mólecules even smaller than *nanoparticles*.

The big difference with real, eutactic, nanotechnology is in the opposite direction. Current technology cannot produce products and energy without releasing floods of molecular-level junk into the environment. If you want to produce billions of chemically reactive nanoparticles, just build a fire in your fireplace, or drive your car, or cook something. If you can smell it, it is releasing molecules that fill the air.

Eutactic processes change that. Every molecule is accounted for. Rather than dump odd chemicals and particles into the environment, they can be broken down to small constituents and used to build the next thing. Instead of producing carbon dioxide, for example, the nanofactory would release pure oxygen, and save the carbon to make diamond. The only combustion product normally released would be water.

## OTHER DANGERS

Suppose you were transported back to 1900 and asked to explain to people what dangers were posed by the motor car. If you said pollution. You had be laughed at. You will understand why if you have ever ridden a train with a coal-burning steam locomotive. Steam engines of the day produced a lot more noxious smoke than gasoline engines. A report of deadly crashes might find more favour—people were scared of these unusual mechanic monstrosities. But you'd still be wrong: cars have saved a lot more lives than they have taken, and the average life-span has increased by more than twenty years since 1900. Traffic jams? Hard to imagine in a country with their acres per person. Fragmentation of the traditional extended family and dissolution of sexual mores? Excuse me, sir, are we speaking the same language?

The fact is that if you had to wait to die in a car crash, you had live in average of 6,700 years. Traffic jams are a nuisance, but except in the most densely populated areas, there is no great demand for alternatives such as public transportation. In my own experience the greatest problem with cars in cities is parking.

The social effects are perhaps somewhat closer to the mark. In fact, the twentieth century has seen changes in our life-styles that would have horrified our great-grandfathers. But are we the worse for them? In some cases, perhaps; in others, probably not. In most cases there are compensations—the changes have made things different rather than particular better or worse. What is more, the changes came about from a confluence of causes, including cars, two world wars, radio and TV, the pill, the jet airliner, the computer, and the Internet.

The printing press has been blamed for the Reformation and some hundreds of years of religious wars. This could obviously have been handled better, but these were the growing pains for moving from a largely illiterate culture to a literate one. It does give us a clue as to the real dangers that lie ahead, though. Entrenched power will react strongly to change, even if the change ultimately benefit most people.

Imagine the politician or corporate or labour leader, who became possessed of a device that would allow him to put you in a chair, clap a helmet over your head, and five minutes later have you step out, a fanatical devotee of his, anxious to grab others and hold them down in the same chair. This is not something that anyone has any idea how to build right now, even given full-fledged nonotechnology. But biological knowledge is advancing as fast as any other realm of science, and faster than most. I had find it surprising if by the end of the century; something like this were not possible.

Suppose a great nation decides to do the same with its army, promising to set them back to normal after the "emergency" is over.

Suppose every citizen is put through a milder form of reprogram-flung "to *prevent terrorism.*"

*Nanodrones*, described above as an *antiterrorism weapon*, could easily be misused as an assassination weapon. Nanodrones are early, not advanced nanotech, probably Stage III or thereabouts. Crude, only by comparison, microdrones can be and are built today. These have cameras only; as far as the author is aware. Even they threaten significant social change following to ubiquitous surveillance.

Being constantly spied on is one thing. Being subject to termination with extreme prejudice on someone is whim is another. Nanodrones could be quite difficult to stop, and virtually impossible to trace. If they became generally available, it might be a bad time to be a lawyer. More seriously, note that the United States already has used its (full-sized) drones for assassination, and that governments as a class, over the twentieth century; killed tens of millions of their own citizens and at least as many of the citizens of other countries. Even if governments can be forced to open their operations to public scrutiny and oversight, other organizations will make obtaining nanodrones a high priority. It will be a bad time to be Jimmy Hoffa.

Note that like the other threats, nanodrones are not nearly so formidable to a population that is equipped with nanotechnology. A drone would be useless against someone living in a nanoskin or surrounded by Utility Fog, for example. The danger is that governments might, in a protectionist panic, limit access to countermeasures. This could easily be a slippery slope to a situation with nanotech-equipped officials and powerless, unprotected commoners. And government of the people, by the people, for the people would perish from the Earth.

## Living in Interesting Times

The real dangers that will come with nonotechnology are not the end of things that play well on a movie screen or can be explained in sound bites. They are not dangers from the technology itself, but the effects of shortsightness and greed in the face of a revolution in human affairs. Every major, life-changing development has been accompanied by some strife. Much of the strife has been, in retrospect, unnecessary; but it never seemed that way at the time. In many cases, as in the Reformation, people didn't really understand what they were fighting over until long after the dust had settled.

One of the best ways to prevent, or at least minimize, strife as nanotechnology is developed is for there to be a broad understanding of what benefits it can bring and what the dangers really are. If nanotech remains the dimly understood magic of a powerful few, trouble lies ahead. But if people widely understand how personal manufacturing technology can bring independence and a comfortable life-style to everyone, and the way is made clear for this to happen, cooperation—and sanity—may just prevail.

# 11

# BASICS OF NANOTECHNOLOGY

*Nanoscale science* and engineering most likely will produce the strategic technology breakthroughs of tomorrow. our ability to work at the molecular level, atom by atom, to create something new, something we can manufacture from the "*bottom up*," opens tip huge vistas for many of us. Breakthroughs could bring us nano-structured metals; ceramics and pomers at exact shapes without machining; nano-coatings for cutting tools and electronic, chemical, and structural applications; nano-instrumentation for in micro spacecraft avionics; nano-structured sensors and nanoelectronics for embedded health management systems in aircraft structures and systems; and thermal barrier and wear-resistant nano-structured coatings. There are huge possibilities for nanotechnology applications. This technology may be the key that turns the dream of space exploration into reality.

Of course, all of this assumes a level of maturity in nano technology that remains to be achieved. Meanwhile, we have been watching the progress in this field and even using some of the concepts and early results for conducting some of our own early-stage projects.

In a program called "*Structural Amorphous Metals*," for in stance, Boeing is helping develop a process for controlling the nanostructure of aluminum to make it as strong as titanium while remaining light as aluminum—to improve vehicle bodies and structures.

In another project, we are using *nanomaterials* to improve the peel and delamination resistance properties of composites in order to eliminate the need for fasteners.

In "*Optical Beam Steering*" program, nanoscale optical materials are being applied to produce ultra-small laser communications devices.

In addition to being able to operate in military environments, these devices are also much lighter, more energy efficient, cooler, more powerful and more reliable than previous devices.

Reaping the benefits from this science is a long-term journey that will be challenging for all those involved. And, while some of the near-term applications are encouraging, it is the long-range possibilities that have the potential to change our world at the most fundamental level and lead to an unparalleled economic revolution. Because we are at just the beginning of this potential revolution, it is hoped that a high commitment to this journey will be a critical factor in success fully achieving the potential of nanotechnology. To this end, it is advisable to take the necessary steps to make the dreams of nanotechnology come true. It is up to us to make the 21st Century the nano-age!

What is *nanotechnology*? By definition, it is the application of nano science to useful devices. *Nanoscience* is in turn the science that deals with objects with at least one dimension between one and one hundred nanometers in length, a size range called the *nanoscale*. A *nanometer* is one one-billionth of a material which is pretty close to one one-billionth of a yard. For comparison, a human hair is approximately 50,000 nanometers across, and a nanometer is as much smaller than a football as a football is smaller than distance from the earth to the moon. Anything small enough to be measured in nanometers is much too small to be seen with the naked eye.

So if *nanoscience* is the science of dealing with objects whose size is almost inconceivably small, why does it get so much hype, and why is it so important for national defense and homeland security? The first reason is that nanoscale objects are not just small, they are a special kind of small. Individual atoms are around one-fifth of a nanometer. The size of almost all molecules from alcohol to sugar to caffeine lies within the nanoscale, because it is the smallest level at which functional matter can exist—anything smaller is just a minute speck of vapour. Material designed at the nanoscale can there fore be designed with molecular precision. This means that, through nanotechnology, we can make materials whose amazing properties can be defined in absolute terms: "This is not only the strongest material ever made, this is the strongest material it will ever be possible to make."

There is another aspect of the nanoscale that makes it important. It is the scale at which the quirky quantum mechanical properties of matter and its more familiar mechanical properties (such as hardness, temperature, and melting point) meet. At the nanoscale it is possible to take advantage of both sets of proper ties, and this allows us to do things that simply cannot be done any other way. This duality is essential to the basic processes of life, which is why nature builds at the nanoscale. Most fundamental biological structures including DNA, proteins, and enzymes do their jobs at the nanoscale (cells are much larger), working molecule by molecule to build the macroscopic structures we call leaves, tadpoles, weevils, and humans.

The properties of smallness, then, are what make nano science and nanotechnology so important. By creating structures whose size is the same as individual molecules or collections of molecules, we can create devices with new and unique properties that solve many of the most intractable problems. *Nanofabrication* techniques let us make bulk quantities of materials designed at the nanoscale and bring these properties into the macro world. The potential applications of nanotechnology run the length and breadth of society and of industry, but the ones with which we will be most concerned in this book are those involving security.

## Misconceptions

There are a few common misconceptions about what nano technology is. For the most part, these misconceptions arise from the portrayal of nanotechnology in science fiction, and they have become so fixed in the public attention that they distract from the genuine promise of the nanoscale. The first of these misconceptions is the concept of a molecular assembler. The other is the concept of "*gray goo.*"

Let's start by examining molecular assemblers. This idea is derived loosely from the writing of Eric Drexler, founder of the Foresight Institute and coiner of the term "*nanotechnology.*" The idea is that a device could be created to place individual atoms together to form any arbitrary nanostructure desired. As the device assembled these structures, they could in turn be placed together to make macroscopic amounts of any given material or manufactured item from perfect rubies to kitchen sinks to armored tanks. If they existed, these assemblers could be carried along in battle to repair damage and all repairs would be absolutely as good as new. On the surface, the idea sounds futuristic but plausible, given nanotechnology's ability to work at the molecular scale. Clearly the benefits of such a device would be enormous. A factory of molecular assemblers could make any item without flaws, and it could go into action on the battlefield to do everything from healing wounds to disassembling enemies' equipment.

After a bit of closer scrutiny, though, the cracks in this argument begin to show. There are a number of compelling reasons why molecular assemblers are either impossible or are at best in our distant future, and it's worth looking at a few of them in order to read sci-fi without nightmares.

One way to understand the challenges of molecular assemblers is by thinking about the *Lego robot* example. Suppose that you want to build a robot out of Legos whose job it is to build other things out of Legos. The only way for your robot to reliably move a Lego piece would be to attach to it in the traditional Lego brick fashion, but then it could not easily let go. (In essence, it has sticky fingers—no fair reaching in to remove the Lego from the manipulator.) Your robot would also have trouble placing additional parts in tight places, because the bricks making it up are too big (for, though we do not wish to be unkind, it has fat fingers). You have no command and control device to tell your robot where to put the next Lego (you can't just connect a computer, because even a simple circuit is a million times bigger than your robot), and you have no power source to make your robot go.

All in all, then, while your Lego robot is probably very cute, it is hopelessly unsuitable for building more things out of Legos, and all of its problems are analogous to those faced at the molecular scale. Sticky fingers that cannot release parts are the result of inconvenient chemical interactions. Fat fingers that cannot operate in tight spaces, the power supply issue, and a lack of control circuitry arise because the nano scale is the ultimate scale of miniaturization and can't accommodate macro scale devices like batteries. Even Nature's best assembler, the cell, can only accommodate all this machinery by being much larger than the nanoscale.

If one Lego robot cannot do the job, might we get some where with an army of them? No. Without a power source, communications, or control circuitry, these robots could not work in parallel. There would be no way to coordinate action analogously at the nanoscale. This means that a single assembler would need to act to build up your materials atom by atom. Even assembling at the rate of millions of atoms per second, it could take billions of years to generate even a few grams of target material.

Finally, even beyond these problems, there is the issue of molecular stability Molecules resemble card houses in some ways. Once they are completely fitted together, they are stable, but there are many stages during their construction at which they will just fall over if not supported. There are design rules arising from quantum mechanics—only some card house structures are stable. For example, carbon atoms can bond to one, two, three, or four other carbons, but not to five or more. Working alone, a single manipulator placing atoms in a structure could not provide this support. This could result in any number of problems; to pick a simple one, a robot trying to make water molecules ($H_2O$) from supplies of hydrogen and oxygen by mechanical assembly (using manipulator arms rather than allowing chemical reactions to happen spontaneously) might instead produce little clouds of hydrogen gas ($H_2$) that Could go wafting off before the final oxygen could be added.

Making particular *nanostructures* is of course totally practical. As we discussed extensively in Nanotechnology: A Gentle Introduction to the Next Big Idea, both synthesis of molecules and fabrication of nanostructures are quite advanced areas of science and technology. Almost any pharmaceutical plant makes nano-sized molecules, from Tylenol to Vancomycin to Viagra. The process includes chemical reactions and physical separations like crystallization, with no hint of nanoscale con rolled robots.

Some of the newest fabrication methods can indeed prepare a wide variety of individual nanostructures. For a beautiful example, Don Eigler's group has made the IBM initials in letters a few atoms tall. But the tools used are macro scale probe tips, and the structures are stabilized by sitting on microscopic metal surfaces. Even then, manufacturing nanostructures this way a single machine can yield only .000000000000001 ounces of material per year.

So, while the idea of molecular assemblers is great, practical considerations get in the way. We will have to find other means of making self-healing materials for use in battle and this book will suggest a few of them.

The other common misconception about nanotechnology is the nightmare "*gray goo*" scenario described by people including Bill Joy at Sun Microsystems Prince Charles (who probably got his ideas by divine right), Chris Carter, and Michael Crichton. The idea here is that little nano machines (described as gray goo) could invade the blood stream of people or the computer systems of starships, communicate with one another, self-replicate, and possibly colonize our bodies, control our minds, or simply consume or destroy us.

While the gray goo scenario and its derivatives are certainly frightening ones (and great bases for good science fiction novels), they are emphatically not what nanotechnology and nanoscience are about. Self-replication is one of the secrets of life, but we know enough about it to realize that simple nano structures, of the sort that nanotechnology will try to make, can't do it. *Cognitive intelligence* of the kind envisioned in this scenario is also impossible at the nanoscale. Human-level intelligence requires extraordinarily complex machinery and intricate communication channels. The average human brain weighs about three pounds. It is certainly not nano sized. Even the functional components of human brains, nerve cells, are much larger than the nanoscale and need to be fed by an equally complex system of blood vessels and protected by bones and membranes. In short, the gray goo scenario is far fetched for many of the same reasons that molecular assemblers are. Functions this complex simply don't work at a scale this small, and even if nanobots were the size of bacteria or viruses, we have the human immune system and a host of other technologies for defending ourselves against them.

## Offshoots

The reason we spent so much time examining the feasibility of molecular assemblers is to demonstrate that nanotechnology's primary importance does not lie in finding new ways to make existing products, but rather in the opportunities it offers to make entirely new things that can only be created at the nanoscale. The scope of such creations is vast. In the con text of defense and homeland security we are most interested in seven major areas: materials, sensors, biomedical nanostructures, energy, electronics, optics, and fabrication. Each of these major areas will be examined on the following pages.

### Materials

*Materials science*, which deals with the properties and struc ture of materials, is the youngest branch of engineering. Materials designed at the nanoscale have remarkable properties because their exact molecular or atomic structures can be precisely tweaked. A natural example of this is the difference between diamond and graphite dust. Both are made up of 100% carbon atoms, and the only difference between the diamond in an earring and the black stuff in a pencil is molecular structure. *Nanotechnology* is concerned with applying this principle to man-made objects. Examples of these *nanoscale* designed materials include ultrathin molecular coatings that turn ordinary cotton cloth into water-repellent, stain-resistant fabric and ultrahard coatings that make materials resistant to scratching and abrasion. Other nano materials include zeolite molecular sieves, whose nanoscale pores and channels can be used for applications such as petroleum refining, oxygen separation from air and water softening, or piezoelectric materials, which can change their dimensions when electrical current is passed through them.

Perhaps the most remarkable example of a nanomaterial is a new member of the carbon family called a *nanotube*. Nanotubes were discovered in the 1990s and they come in several varieties, but the most researched are single-walled nanotubes that look like nano scale soda straws. Like diamond and graphite, *nanotubes* are made out of only the element carbon. Their diameters can range from less than one nanometer to near 20 nanometers, and they can be best envisioned by imagining chicken wire, wrapped up onto itself and then welded together. Each three- wire crossing point within the chicken wire is analogous to a carbon atom in a nanotube, but of course the nanotube roll is ten million times smaller. Nanotubes are the strongest, most conductive, and stiffest materials ever made (over 60 times stronger than steel by some estimations), and they are designed at the nanoscale. Some scientists have estimated that a nanotube cable the size of a human hair could suspend a locomotive or reach from the Earth to the moon while bearing its own weight.

Materials designed at the nano scale are often called *smart materials* since they are custom designed for a particular purpose. They can also be made to react to outside stimuli, as in the case of self-tinting window glass which darkens in direct sunlight. These are called *dynamic smart materials* and will be key to many military applications.

**Sensors**

A *sensor* is any device that gives a recognizable signal in response to the condition or thing it is designed to detect. This signal can be any discernable change in the properties of the sensor such as the colour change of litmus paper in the presence of acid, the shriek of a burglar alarm set off by an intruder, or the vibration of a cell phone sensing a call. In nature, noses, eyes, and other organs act as *sensors*.

*Nanoscale sensors* are generally designed to form a weak chemical bond to the substance of whatever is to be sensed, and then to change their properties in response (that might be a colour change or a change in conductivity, fluorescence, or weight). For sensing biological species such as toxic organisms or bioterrorism agents, sensors usually work by specifically binding to the target DNA of the given biological agent.

Sensors could be used to detect such substances as oxygen, anthrax toxin, carbon monoxide, or nerve agents. Development of sensors is one of the major applications of nanotechnology in global security.

**Biomedical Nanostructures**

*Biomedical nanostructures* are designed to interact with particular molecular or larger scale structures within a biological organism, particularly within people. Simple biomedical nanostructures are used in drug delivery to encapsulate small amounts of a drug, thus speeding its dissolution within the body and its availability at a local site where remediation of pain or disease is necessary. *Magnetic nanomaterials* can also be used to aid in drug delivery, for they can be attached to a drug molecule and then directed externally, by magnetic fields, to different parts of the body. Applications of biomedical nanostructures in the treatment of disease include adhesive materials for skin grafts or bandages, oxygen barrier structures for burn therapy, and tiny metal or semiconductor tags that can be attached to particular proteins or medium-size molecules and used, with microscopes, to observe the actual biological function or structure of these molecules. Such bio-nanostructures will be crucial in assuring effective home land and military defense.

**Energy**

In addition to homeland defense, global climate change and the energy supply are major issues in national and international security. Nanostructures play key roles in scenarios for reducing climate change through reducing the burning of carbon-containing materials, and for making the energy supply entirely independent of hydrocarbons such as petroleum. Particular nanostructure applications in energy include the use of titania nanoparticles (titania is the white pigment in house paint) to capture electricity from the sun. They include fully integrated nanoscale fuel cell structures for computer backup power and portable energy, ultrathin separation layers for advanced batteries, and transparent electrodes for energy efficient lighting using so-called light emitting diode structures. Finally they include nanoporous carbon electrodes for advanced, high energy, and high power battery structures.

**Computers**

*Modern computers* are inexpensive and effective because their manufacture actually involves nanoscale structures. Ever since the transistor, which is still the fundamental element of a computer, was developed at Bell Laboratories in the 1940s, it has been getting smaller. The length of the current transistors is roughly 100 nanometers, and therefore these transistors exist at the nanoscale. Transistor materials include silicon, silicon nitride, silicon dioxide, gallium arsenide, and other exotic semiconductors. But despite existing at the nanoscale, semiconductor electronics do not yet take advantage of any of the unique properties of the nanoscale. They simply represent the continued miniaturization of components. For this reason, modern semiconductor manufacture is not considered by many to be true nanotechnology. Unlike nanotechnology, it is a technology that is reaching maturity and approaching its fundamental limits. Nanotechnology can open the doors to new possibilities.

Some nanotechnological approaches to the problem of continuing to shrink circuits and make them more efficient change the rules at a very basic level. For example, scientists are now able to build circuits with individual molecules as circuit elements. Indeed, it was Richard Feynman's speech in 1959 entitled "*There's Plenty of Room at the Bottom*" that began to turn scientific thought toward nanoscience. In that speech, Feynman talked specifically about electronics applications of nanostructures in computing and in memory.

Not all electronic applications of nanostructures are involved in information technology, though. Nanometer-thick conducting wires can also be used to dissipate static electricity. In addition, both light emitting diode structures and candidates for all-optical computing utilize nanoscale structures for their function. Given the crucial importance of information and electronics in defense at all levels, nanoelectronics will be a featured component of future security scenarios.

**Nanoscale Optics**

*Optics* is the study of light, but by nanoscale optics we generally mean the interaction of light with nanostructures. Applications of nano-optics include control of colour by control of nanoparticle size (this is called the *quantum* size effect, and it is one of the most striking results in nanoscale science: pure gold can range in colour from green to dark red, depending on the size of the nanoparticles). Materials with controlled fluorescence can also be designed at the nanoscale, and once again their properties, including their colours, can be tuned. Some of the other applications mentioned above, including light emitting diode structures and *photovoltaic* (conversion of sunlight to electricity) devices, are also applications of nanostructures in the general areas of optics. Finally, active camouflage is a very promising nanoscale-based soldier protection scheme.

**Nanolithography**

*Fabrication* is perhaps not so much an application as a tool, but it is of major importance: if we are to understand and study nanostructures, we need to be able to make them. For example, *NanoInk*, a Chicago-based startup, uses nanoscale structures as pens to write very narrow (a few nanometers thick) ink-like lines of specific molecules on surfaces. This is called *dip-pen nanolithography*. It is also possible to employ spheres of nanoscale dimension laid out on a surface to act like a mask for spray painting, resulting in a pattern of hollow tri angles (made of molecules that behave like the paint) of nanoscale dimension isolated on the surface. This is called *nanosphere liftoff lithography*. Continuing the analogy between macroscale and nanoscale patterning, we can make stamp pads of a soft plastic material called PDMS, that can be used to stamp out arbitrary patterns with molecular "inks" on appropriate surfaces.

Since unlike charges attract, we can use the attraction between positive and negative charges to make layers only one molecule thick. This is done by first laying down a layer of positively charged molecules, then a second layer of negatively charged molecules, then the third layer of the positively charged molecules again, and so on. This fabrication method can be used to design very thick (macroscopic) structures that are homogeneous on the macroscale and layered at the nanoscale.

Although these capabilities have been developed in academic labs by the groups of Chad Mirkin, Rick van Duyne, George Whitesides and Mike Rubner, respectively, they are now being pursued actively by startups, often with defense applications in mind.

As already indicated, nearly all of the major devices used in modern computational structures, such as desktop, laptop, or workstation computers, are based on some rather simple structures. One reason it costs so many billions of dollars to build a fabrication plant for computers is that as the components shrink to nanoscale dimension, unusual forces such as the capillary force (which also makes wet sand castles much stronger than dry ones) and other molecular binding interactions can be the major directors of structure. Learning to use the resulting structures to do computations is a major challenge. Feynman first suggested that if we are to continue to make computers smaller and smaller, we will eventually have to stop making them by carving up large slabs of silicon and start building them the way nature builds us—by assembly of molecules starting from the bottom and getting bigger. This kind of bottom-up technology will characterize much of the nanotechnological world.

## Nanoscience

*Nanoscience/nanotechnology* is important and exciting not simply because of its small dimensions. "*Small is beautiful*" is a cute phrase, but unless there are real advantages in using nanoscale structures, the trouble is not worth it. It is important and exciting because there are unique capabilities offered by working at the nanoscale. Perhaps the most important of these is the ability to control the interaction of nano devices with other systems: with molecules to be sensed, with sunlight to be converted to electricity, with pollutants or toxins to be kept out, with viruses or bacteria to be destroyed, with injured or diseased biological structures to be repaired, or with sound to be amplified in replacing hearing loss. In each of these cases and hundreds more, working at the nanoscale has unique capabilities.

These unique capabilities, as we have said before, come from the fact that in nature, certain properties occur only at the nanoscale. These include so-called *molecular recognition*—the ability of molecules to bind specifically to other molecules in unique and understandable ways. This capability is what makes biology work, and therefore if we are going to interact in a positive way with biological systems, repairing or replacing or augmenting or reducing difficulties, the agents that we use must be nanosized. Similarly, if we are to control where molecules go (everything from wicking away perspiration to controlling frost heaves and cracking on road surfaces), the control must be at the nanoscale. For example, the flickering metal *nanodots* that are used to label biological molecules with light emitting properties only work when those dots are *nanosized*. The photovoltaic capture of sunlight using titania cannot be done with bulk particles, but only with nanosized grains.

The next question facing nanotechnology is not "Can these nanotasks be done?" but rather "Can they be done efficiently, cheaply, and in an integrated fashion?" These are challenging issues. Although the nanoscience revolution is in full swing, the issues of reliability, cost capability, manufacturability, utility, compatibility, and recyclability have not yet been extensively investigated. Still, the elegance of the small is striking: the enormous complications and high costs associated with making something as complex as a jetliner, an automobile, or a milling machine cannot occur at the nanoscale simply because the structures are too small. As we learn how to build and manipulate nanostructures better, nanotechnology should become inexpensive and widespread, providing unique advantages in everything from cosmetics to security

Although individual nanostructures are too small to see with the eye or feel with the finger, extended nanostructures are visible, tangible, and important. Manufacturing capabilities for these structures can sometimes be developed very quickly. For example, two startup companies in the Chicago area have the capability to manufacture extended nanostructures for materials applications. One, Advanced Diamond Technologies, uses technology originally developed at Argonne National Laboratory to coat structures with nanoscale diamond. This diamond coating is, like diamond itself, very hard, chemically resistant, and atmospherically stable. Because the coatings are so thin, the only substantial changes to the original material are its new stability, hardness, and wear. Another company, *Nanophase*, manufactures powders of metal oxides with nanoscale dimensions. These materials can be used for surface coatings and for catalytic applications, and *Nanophase* can manufacture them by the pound or by the ton.

Perhaps the most important aspect of nanoscience and nanotechnology is the ability to respond to what might be called grand challenges. These are major problems such as diagnosing particular forms of cancer, stopping corrosion on metal bridges, providing early warning of heart malfunctions, developing environmentally friendly and significant new energy sources, providing total assurance of food safety, producing reliable long-term storage of information, and so on. It is in addressing these grand challenges that the greatest potential of the nano sciences to contribute to the betterment and safety of society can be uncovered.

# 12

# Societal Implications of Nanotechnology

Despite the fact that "*nanotechnology*" is still in its infancy—is even, arguably, still prenatal—indeed, despite ongoing disagreements about how "*nanotechnology*" ought to be defined, knowledgeable people have converged around the notion that, whatever nanotechnology is, and whatever it will become, its implications for society are going to be transformational, perhaps radically so, in social realms as diverse as privacy, workforce, security, health, and human cognition. One apparent consequence of this convergence is the commitment by the U.S. federal government to fund not just nanoscale science and engineering (NSE) research, but also research on the societal implications of NSE.

In the present text there are three brief narratives to illustrate the surprisingly independent evolution of (i) NSE research, (ii) speculations and concerns about the implications of nanotechnology, and (iii) government commitment to supporting research on the societal implications of nanotechnology. Conspicuously absent from these stories is the influence of several decades of scholarship on the interactions of science, technology, and society.

The community of science studies and science policy scholars seem to have engaged with the challenges of nanotechnology only when stimulated by the appearance of federal research funding starting in about 2001. Thus, they did not materially participate in the framing of public discourse about nanotechnology, or in the design of research programs to study the social implications of nanotechnology. In their absence—perhaps due to their absence—a policy experiment was implemented that may permit this same community to play a newly effective role in the governance of science and technology.

## NANO NARRATIVES

### How Nanotechnology Began to Get Big

The canonical story of the origins of nanotechnology goes something like this: In 1959, Richard Feynman gave a speech at the annual meeting of the American Physical Society called "*There's Plenty of Room at the Bottom*," in which he predicted that physicists would eventually be able to manipulate matter at the molecular or even atomic scale, and would thus usher in a new technological revolution. "It doesn't cost anything for materials. There may develop billion of tiny factories, models of each other, which are manufacturing simultaneously, stamping parts, and so on. As we go down in size, there are a number of interesting problems that arise... But it is not considered as the final question as to whether, ultimately—in the great future—we can arrange the atoms the way we want; the very atoms, all the way down!".

The tools to begin to pursue Feynman's playful predictions started to come on line in the coming decades. In 1980, IBM scientists used a scanning tunneling microscope to directly image individual atoms for the first time. The development of the atomic force microscope in the mid 1980s further advanced imaging capabilities, and in 1990, again at IBM, scientists actually manipulated individual Xenon atoms to write their company logo (NSTC, 1999; 2000). A giant step had been taken toward confirming Feynman's assertion that it should be possible to print the entire *Encyclopedia Britannica* on the head of a pin. Meanwhile, physicists Harold Kroto, Richard Smalley, and Robert Curl discovered in 1985 that carbon exposed to high temperatures could form spherical molecules, later dubbed "buckyballs," which rapidly led to the discovery of numerous, similar carbonbased molecules characterized by both great chemical stability, and great physical strength. *Science* magazine named buckyballs the "molecule of the year" in 1991. Kroto, Smalley, and Curl shared the 1996 Nobel Prize in physics.

Investment in nanoscale science and engineering (NSE) continued to expand. The National Science Foundation (NSF) began it's first program devoted exclusively to NSE in 1991, funded at about six million dollars. In 1998, the government organized the Interagency Working Group on Nanotechnology (IWGN), whose work led, two years later, to the initiation of the multi-agency National Nanotechnology Initiative (NNI), funded at $270 million; by 2004 the investment had increased to $961 million. NSF's 2004 solicitation for NSE research proposals provides an update on Feynman's original vision; still prospective in terms of actual outcomes, it is nonetheless inaugural in pronouncing that the revolution has arrived: "The nanometer (one billionths of a meter) is a magical point on the dimensional scale. Nanostructures are at the confluence of the smallest human-made devices and the largest molecules of living systems... A revolution has begun in science, engineering and technology, based on the ability to organize, characterize, and manipulate matter systematically at the nanoscale. Far-reaching outcomes for the 21st century are envisioned in both scientific knowledge and a wide range of technologies in most industries, healthcare, conservation of materials and energy, biology, environment, and education."

Scientific productivity grew apace. Starting in the early 1990s, the prefix "nano" began increasingly to appear in the titles of scientific journal articles and scientific grant proposals. To some extent this trend almost certainly reflects the opportunism of scientists relabeling existing research activities to take advantage of the latest funding fad. But the trend also signaled the effects of technological and conceptual advances that increasingly allowed new types of research on materials and processes at the nanoscale, and the synergies of such new opportunities with increased availability of research funds, especially from the government.

From 1985 to 1990, the annual number of publications in the Science Citation Index (SCI) that included the prefix "*nano*" in the title hovered between 200 and 400. Between 1990 and 1991, the number jumped from 378 to 1677, most likely reflecting a response to the growing programmatic focus of federal funders. What appears to be exponential growth in publications continues: in 2003, SCI lists 23,015 papers with the "nano" prefix in the title. In parallel, the number of NSF grants with the prefix "nano" in the title increased from the low 10s in the late 1980s and early 1990s, to almost 600 in 2003. The first journal devoted exclusively to nanoscale science and engineering was launched in 1990; two more journals were started later in the decades, and an additional four were launched between 2001 and 2003.

**From Weird to Wired**

The word "*nanotechnology*" appears to have been coined by a Japanese engineer in 1974, although it entered the public lexicon through

Eric K. Drexler's 1986 book *Engines of Creation*. Drexler's vision for the future of nanotechnology was largely an extrapolation of Feynman's original idea, taken to its logical extreme: "Molecules will be assembled like the components of an erector set, and well-bonded parts will stay put. Just as ordinary tools can build ordinary machines from parts, so molecular tools will bond molecules together to make tiny gears, motors, levers, and casings, and assemble them to make complex machines." The crucial attribute of nanotechnology, in Drexler's vision, was the capacity for self-assembly, in order to create necessary efficiencies of time, energy, and scale. Self-assembly "will let us build almost anything that the laws of nature allow to exist. In particular, they will let us build almost anything we can design - including more assemblers... Assemblers will open a world of new technologies. Advances in the technologies of medicine, space, computation, and production - and warfare - all depend on our ability to arrange atoms. With assemblers, we will be able to remake our world or destroy it. So at this point it seems wise to step back and look at the prospect as clearly as we can, so we can be sure that assemblers and nanotechnology are not a mere futurological mirage."

As an MIT-trained engineer, and standing on Feynman's shoulders, Drexler possessed technical legitimacy, one consequence of which was a brief review of *Engines of Creation* in the *New York Times*, which noted Drexler's "unembarrassed faith in progress through technology," and voiced general skepticism about his vision of "*molecular manufacturing*." Reviewing *Engines* in *Technology Review*, the noted robotics engineer Hans Moravec (1986) mostly poked fun at Drexler's "*absurdly anthropocentric*" nano-utopia, yet also asserted "atomic scale construction is not just possible but inevitable in the foreseeable future."

It would be nearly 15 years before Drexler began to get broad attention, but science fiction writers were meanwhile spinning out visions of what molecular manipulation and self-replication might imply. Moravec's 1986 review of *Engines* included the idea that "medical nanobes [might] rebuild you from the inside out in their own image." Yet a year earlier, Greg Bear's novel *Blood Music* (1985) had spun out precisely this scenario, where a discredited scientist injects himself with self-replicating "nanites" that take over and "improve" his body, then escape and do the same to the rest of humanity. In the same year, Paul Preuss's novel *Human Error* (1985) speculates about self-replicating hybrid bio-nano machines that can enhance the performance of their human hosts.

Between 1985 and 2000, at least 37 science fiction novels were published that spun out a variety of nano-enhanced futures. Many of these were concerned with exploring not just the technological implications of nanotechnology, but also the social dilemmas and consequences that might ensue. As Neal Stephenson imagines in *The Diamond Age* (1995): "Now nanotechnology had made nearly anything possible, and so the cultural role in deciding what should be done with it had become far more important than imaging what could be done with it."

Despite this activity outside the formal boundaries of technoscience, public interest in nanotechnology appears to have been modest throughout the 1990s. Mentions of the word "*nanotechnology*" in popular print media included in the Lexis-Nexis database rose very gradually and arithmetically, from 10s per year to a total of 183 in 1999. Most of these mentions were magazine and newspaper coverage of the latest scientific breakthroughs and technological possibilities. Nanotechnology was still the stuff of *techno nerds*.

Then something changed. Media mentions of "*nanotechnology*" more than doubled in 2000, to 423, and by 2003 had exceeded 1750. Michael Crichton's nanotechcatastrophe book *Prey* (2002) became a national bestseller. *Scientific American*, a magazine about science but for general audiences, ran seven articles with the prefix "nano" in the title between 1993 and 2000; the number jumped to 54 between 2001 and 2004. Similarly, *Technology Review* published 44 articles that included the keyword "nanotechnology" between 1997 and 2000; over the next four years the number more than quadrupled, to 183.

Why did nanotechnology so suddenly become a mote in the public eye, a buzzword that epitomized the rapid advance of science and innovation? Certainly the rapid growth of research interest and productivity in the NSE field made it ripe for media interest, and the launching of the NNI in 2000, which was widely covered in major print media, correlates with the rapid expansion of media coverage. Much of this coverage continued to focus on the latest breakthroughs at specific laboratories. We speculate that, among other factors, as universities strove for a piece of the expanding NSE budgetary largesse, they promoted their own NSE research activities more aggressively to the media.

Yet a stimuli of perhaps equal importance was the now-famous article in *Wired* magazine by Bill Joy, chief scientist at Sun Microsystems, entitled "Why the Future Doesn't Need Us," prognosticating doom for the human species as a result of the emergent power of three converging technologies: nanotechnology, genetic technology, and robotics. Expanding on scenarios already published by Drexler, as well as the ideas of the inventor technological visionary Ray Kurzweil (1998), Joy wrote: "robotics, genetic engineering, and nanotechnology . . . pose a different threat than the technologies that have come before. Specifically, robots, engineered organisms, and nanobots share a dangerous amplifying factor: They can self-replicate. A bomb is blown up only once - but one bot can become many, and quickly get out of control." While Drexler had devoted a chapter to the possible dangers of self-replication, he had also considered it to be a manageable problem. Joy saw it as intrinsically uncontrollable. His solution?: "relinquishment: to limit development of the technologies that are too dangerous, by limiting our pursuit of certain kinds of knowledge." Joy's position as one of the chief architects of the world's high technology information infrastructure meant that he could not be dismissed as a fringe voice or Luddite, The contrast between the "gee whiz" utopian breathlessness of the scientific promoters of nanotechnology, and the "oh my god" catastrophism of Joy's vision, created a tension ideally suited for journalistic treatment. His article received broad coverage, and thus legitimation, in the mainstream media, including the *Washington Post* and *New York Times*. Nanotechnology was on the radar screen.

**Mud in Joyville**

Rapid increases in the public investment in NSE starting in 2000 had to be explained and justified. The inauguration of the NNI was accompanied by a promotional brochure aimed at non-technical audiences, entitled "Nanotechnology: Shaping the World Atom by Atom", which proclaimed nanotechnology as "a likely launch pad to a new technological era because it focuses on perhaps the final engineering scales people have yet to master." "If present trends in nanoscience and nanotechnology continue, most aspects of everyday life are subject to change." "The total societal impact of nanotechnology is expected to be much greater than that of the silicon integrated circuit because it is applicable in many more fields than just electronics." And the ultimate goal of the nanotechnology revolution?: "unprecedented control over the material world."

Such language, which displays an historically oblivious optimism that borders on the quaint, testifies either to a conspicuous isolation of those involved in planning and promoting the NNI from anyone who might have been thinking about the societal complexities of scientific and technological change, or a conscious decision to ignore any such thinking. The publication of "Why the Future Doesn't Need Us" only months after the NNI's unveiling must therefore have been particularly galling to those involved in promoting the initiative.

Indeed, Joy's proposal to "*relinquish*" certain potentially fruitful lines of scientific research is not just unacceptable but literally incomprehensible to most scientists. Advocates of the benefits of nanotechnology thus sought from the outset to discredit the plausibility of Joy's ideas on either the scientific grounds that self-replicating "nanobots," as originally described by Drexler, were impossible, so there was nothing to worry about, or on the grounds that stopping the advance of NSE knowledge was impossible, so we would just have to figure out how to deal with it. NSF's Mikhail Roco, the director of the NNI, and Nobel prize-winner Richard Smalley, the co-discoverer of buckyballs, in particular were known to be highly antagonistic to Drexler's ideas. After the initial flurry of attention devoted to Joy's article, scientific criticism of the Drexler-Joy scenario has been sufficiently effective to keep it out of most mainstream published discussions and accounts of possible social implications of nanotechnology. Although one can only speculate, it seems to us that the high level and persistent energy of scientific critique of Joy and Drexler cannot be rooted in the technical objections to the scenario, but in the unavoidability of Joy's logic: if uncontrollable self-replication of autonomous nanobots is possible, then a strong case can be made that relinquishment of certain lines of investigation is not only rational but sensible. Because relinquishment is unthinkable, self-replication must be deemed impossible.

**Nowhere to be Seen**

As early as 1990 the popular journal *The Futurist* published an article entitled "Nanotechnology and Human Values"; during the 1990s business magazines such as *Forbes* and *Futures* also featured occasional coverage of the implications of nanotechnology, and academic journals such as *Scientometrics* and *Research Policy* began to track scientific and technical trends. As late as 2000, however, nanotechnology was nowhere on the agenda of scholars who study the societal implications of science and technology. Searches on terms such as "nanotechnology" and "nano" in the Social Sciences and Arts and Humanities Citation indices show little if any academic interest in nanotechnology during the 1990s. Mnyusiwalla and others (2003) writing in the technical journal

*Nanotechnology*, noted: "Despite the potential impact of nanotechnology, and the abundance of funds, our research revealed that there is a paucity of serious, published research into the ethical, legal, and social implications of nanotechnology. Conversely, scholarly works that directly confronted the challenges of governing societally transforming technologies took little note of nanotechnology. For example, Richard Sclove's *Democracy and Technology* (1995), an edited volume entitled *Technology and Values*, and the book *Frontiers of Illusion: Science, Technology, and the Politics of Progress* include no mentions of nanotechnology.

Apart from Bill Joy, one other voice urging caution about nanotechnology was the ETC Group, an activist organization that during the 1980s and 90s played a central role in opposition to genetically modified foods, especially to Monsanto's "*terminator*" technologies. A 1999 article by RAFI director Pat Roy Mooney, entitled "The ETC Century: Erosion, Technological Transformation and Corporate Concentration in the 21st Century," included a 10 page summary of the state of nanotechnology research. Unlike Joy's diagnosis, the ETC perspective focused on a combination of equity issues (who would benefit socially and economically from nanotechnology?) and a more traditional risk framework (what are the environmental and health risks associated with nanotechnology?) But overall, the point is that on the eve of the NNI, the community of scholars devoted to understanding the social embeddedness and implications of science and technology were playing no part in the gradually unfolding societal discourse about nanotechnology. Notably, however, in 2001 two academic papers were published that analyzed the role of nanotechnology in science fiction (Johnston, 2001; Miksanek, 2001).

It is not without irony, then, that from its beginnings the NNI, and the U.S. National Science Foundation in particular, proclaimed themselves committed not just to research on nanoscale science and engineering, but to simultaneous research on the "Ethical, Legal, Societal Implications" (ELSI) of nanotechnology to "help us identify potential problems and teach us how to intervene efficiently in the future on measures that may need to be taken" (NSTC, 2000). This commitment, of course, was no invention of the NNI, but rather echoed the decade-old ELSI program of the Human Genome Project, as well as the Human Dimensions of Climate Change initiative of the U.S. Global Change Research Program, and the 1999 recommendations of the President's Information Technology Advisory Committee to include research on "socioeconomic impacts" in the nation's portfolio of information technology research. This was a top-down commitment.

In September 2000, NSF sponsored a two-day workshop on "Societal Implications of Nanoscience and Nanotechnology," a wide-ranging and ill-focused event, which led to a published volume of the same title. The introduction to the volume sets out the rationale for research on the societal implications of nanotechnology: to "boost the NNI's success and help us take advantage of this new technology sooner, better, and with greater confidence." But a later chapter says that "the knowledge gained will help policymakers and the public understand how nanoscience and nanotechnology are advancing, how those advances are being diffused, and *how to make necessary course corrections*." While most of the participants at the workshop and contributors to the volume were concerned primarily with the problem of how to effectively advance nanotechnology and its benefits, several authors did raise questions about the complex outcomes of technologically induced societal transformation.

Over the four years following this conference, NSF funded a small number of grants to academic scholars, ranging in size from about $30,000 to $1.7 million, on the societal implications of nanotechnology, in such diverse areas as the history of the scanning probe microscope, analysis of emerging ethical and risk issues, development of various participatory techniques to enhance public dialogue, and the construction of web-based NSE databases. NSF's total commitment to a broadly construed social implications research agenda during the first four years of the program appears to have been about $10 million; total NSF expenditures on all nanotechnology research during this period were about $750 million; total, multi-agency NNI expenditures by the U.S. Government were about $2.7 billion. Social implications research amounted to less than 0.4 % of the total federal investment in NSE research. By comparison, the ELSI component of the Human Genome Project by law was funded at five percent of total project expenditures.

Through 2004, NSF funding for research on the social implications on nanotechnology was disbursed in a non-strategic manner to diverse universities. The situation may now be changing. In 2004, the U.S. Congress passed the "21st Century Nanotechnology Research and Development Act," which mostly asserted Congressional authority over the funding and coordination of the NNI, but did include specific provisions to ensure "that ethical, legal, environmental, and other appropriate societal concerns" were considered "*during* the development of nanotechnology." Prior to the passage of this legislation, the U.S. Senate Committee on Commerce, Science, and Transportation held a hearing on the subject of nanotechnology which included one witness testifying about the need for social implications research—a professor from the University of South Carolina, the state represented by the Committee's senior Democratic senator. The U.S. House of Representatives Committee on Science, in contrast, convened a hearing specifically entitled "Societal Implications of Nanotechnology." Their witnesses included Ray Kurzweil, the technologist whose optimistic visions of how nanotechnology might evolve nonetheless accepted some of the technical assumptions advanced by Eric Drexler and adopted by Bill Joy; Christine Peterson, who was President of Eric Drexler's Foresight Institute; and the science studies scholar Langdon Winner, a long-time advocate of increased democratic participation in technological decision making. Notably, among the questions that each witness was asked to address was this: "How can research and debate on societal and ethical concerns be integrated into the research and development process, especially into projects funded by the government."

As a consequence of this hearing, language was included in the 21st Century Nanotechnology Research and Development Act that singles out "the potential use of nanotechnology in enhancing human intelligence and in developing artificial intelligence which exceeds human capacity," and includes the requirement for "public input and outreach to be integrated into the Program by the convening of regular and ongoing public discussions, through mechanisms such as citizens' panels, consensus conferences, and educational events, as appropriate"

In response to the passage of the legislation, NSF in August of 2004 initiated a competition for a "Center for Nanotechnology in Society," to be located at a university or consortium of universities, and funded at a level of about $2.6 million per year for five years, with a possible five year extension. This competition is currently underway. The program solicitation includes a much broader range of potential research foci than stipulated in the legislation, such as research to "improve our understanding, e.g., economic implications of innovation; barriers to adoption of nanotechnology in commerce, healthcare, or environmental protection; educational and workforce needs;" but also lists "ethical issues in the selection of research priorities and applications and in the potential to enhance human intelligence and develop artificial intelligence; and public participation and involvement in scientific and technological development and use."

Four years after NSF began to provide support for social science and humanities research on the societal implications of nanotechnology, and on the eve of a new infusion of public funds into this area of study as a result of the new law, a scholarly literature is perhaps beginning to emerge. For example, the February 2004 issue of the *Bulletin of Science, Technology, and Society* was devoted to social implications of nanotechnology. Notably, a survey of the citations in the six articles contained in this volume confirms the absence of a significant prior literature on this issue. This situation will certainly begin to be reversed over the next several years, as researchers begin to report on the results of their federally funded work. But the key point here is that this area of research has been created by a federal funding commitment; it did not arise in response to the societal challenges presented by the emergence of nanotechnology.

## Studies Converged With Common Sense

The December 2004 NNI Strategic Plan (NSTC, 2004) states: "Recognizing that technological innovations can bring both benefits and risks to society, the NNI has made research on and deliberation of [the societal implications of nanotechnology] a priority." What are the mechanisms by which research on the social implications of an area of science and technology are supposed to improve human choices about, and social outcomes related to, that area of endeavor? Who are the constituencies who might use such research, and how might those constituencies act to address what is learned? If, for example, one considers the defunct Office of Technology Assessment (OTA) of the U.S. Congress, the idea, of course, was that specific studies of particular areas of research and innovation would help illuminate the implications of various decision options facing elected representatives. The complexity of Congressional politics meant that the capacity for OTA studies to influence decisions was both highly diluted and highly buffered, but at least the model by which OTA might contribute to decision making was easily understood.

If one considers the ELSI program of the Human Genome Project (HGP), the formula becomes less clear. ELSI research was supposed to "anticipate the social consequences of the projects' research and to develop policies to guide the use of the knowledge it produces." But ELSI research, conducted primarily by academic social scientists and humanists, is functionally and administratively separate from the genomics work that constitutes the core of the HGP. Nor are the results of its research directed at or responding to any decision-making constituency. Moreover, a key tenet of the HGP ELSI program from the outset was that it conducted research on the implications of science emerging from the HGP, but did not address the deeper question of how the HGP science agenda was actually set, or what science actually ought to be done. Nor were there formal mechanisms by which ELSI research could feed back into the science policy making process. Neither the genomics community supported by the HGP, nor the bioethics community who benefited from ELSI funding, sought to change this situation, which in fact protected the autonomy of both.

According to Robert Cook-Deegan's (1994) account of the origins of ELSI, several influential Members of Congress voiced concern that the structure of ELSI—basically, to provide research grants for academic social scientists and humanists—was not likely to prove policy-relevant. As Cook-Deegan explains, efforts by the National Institutes of Health to sponsor a separate policy analysis function that might link ELSI research to policy decision processes were not successful.

The 21st Century Nanotechnology Research and Development Act rediscovers this fundamental defect in the HGP ELSI program. There are no mechanisms to connect policy questions to social implications research agendas, and the processes by which research results are to enhance decision making are not stipulated. However, as Fisher and Mahajan (in review) recognized in their careful analysis of the law, it does demand something new, different, and important: integration of NSE research and social implications research. All NSE research centers are required to "include activities that address societal, ethical, and environmental concerns," and such centers must, "insofar as possible, research on societal, ethical, and environmental concerns with nanotechnology research and development."

When combined with the further requirement that participatory decision mechanisms be included in social implications research activities, this integration of natural and social sciences raises the possibility that nanotechnology research institutions could be structured to build social learning and reflexiveness into the research process, and thereby offer internal guidance for the production of NSE knowledge. This integration, in other words, could make nanotechnology governance a part of the knowledge creation process itself, pushing it "upstream," as it were, into the scientific laboratory, where scientists and engineers are making choices about the types of problems they address, the approaches they use, the outcomes that they seek to pursue. The theoretical and empirical basis for suggesting that such an approach to knowledge production might be societally beneficial builds on the last half-century of social science research into the character of scientific and technological advance. Starting in the 1950s, economists studying the relationship between technological innovation and economic growth began to build a picture of scientific research (including basic science) as embedded in a complex social network. Innovation emerged from the continual interactions of a variety of actors in a variety of institutions, including academic scientists, industrial scientists, research administrators, regulators and policy makers, corporate executives, and consumers.

A second branch of scholarship over the past several decades has revealed the texture of the social embeddedness of research, as elaborated in historical, sociological, cultural, and political approaches. Broader, grounded theory and policy analytic treatments now recognize that any analytical framework for understanding knowledge production systems must be founded on an elucidation of social contexts within which knowledge production is occurring. Science and society, that is, are "co-produced;" they are mutually constituted through a network of actors and institutions in which decisions about science and technology are made.

Is it feasible to operationalize this enormously powerful and well-supported insight in the design of knowledge-producing institutions by making co-production explicit in the knowledge creation process? In particular, through engaging scientists and various publics in discourse about the contexts, meanings and values surrounding nanotechnology (real and imagined), could institutions build a greater capacity for reflexiveness—that is, social learning that can expand the range of conscious choice—in knowledge production—as knowledge is being produced? If one takes seriously the language of the 21st Century Nanotechnology Research and Development Act, it could be interpreted as supporting institutional experimentation to probe the hypothesis that expanding awareness of context and choices in NSE research institutions can be the basis for steering knowledge and knowledge-based innovation toward socially desirable outcomes, and away from undesirable ones.

Yet, as our narratives above are meant to illustrate, social scientists and humanists had little if any engagement with nanotechnology during the 1980s and 1990s, leaving the consideration of societal implications to technologists like Drexler, Kurzweil, and Joy, to activists like Pat Roy Mooney, and to science fiction authors. Nor is there any evidence (although, admittedly, it is difficult to know where such evidence might lie) that science studies and science policy scholars were discontented with the manner in which NSF supported research on the societal implications of nanotechnology, given that such support was disbursed through standard peer-review mechanisms via programs with which scholars were already familiar, and which gave them maximal autonomy.

What, then, are the origins of the policy innovation at the core of the 21st Century Nanotechnology Research and Development Act? Our conversations with the legislative staffers who drafted the bill indicate that they well understood that the integration of social implications and NSE research could allow more informed decision making about the science itself in light of both societal goals and concerns. The staffers also recognized that one of the major failings of HGP ELSI was the lack of such integration. "It's common sense," said one staffer, who also noted that Congress was "divorced from academic politics," meaning that the legislative drafters didn't have to worry about the academic barriers to interdisciplinary research that so often obstruct effective integration—they could simply decree integration as a condition of receiving federal support.

House and Senate approaches diverged on the question of whether social implications work should be an integral part of all NSF-funded NSE research, or whether a major center should be funded as a flagship for such integrated activities. In particular, staff from the Senate Committee on Commerce, Science, and Transportation, whose ranking member was from South Carolina, pushed for the funding of a Center, because they felt that the University of South Carolina, which had already received a major NSF grant for social implications of nanotechnology, would be well positioned to compete successfully for the national center. Local politics was thus a key driver of the institutional innovation at the heart of the Act.

Political necessity was also a key to other important provisions. Science legislation is of a generally low priority in the U.S. Congress, and thus for the most part is only brought to a vote under conditions of unanimous consent. This favours accommodation of minority views, because any one disgruntled Member can thus, in theory, block passage of a bill. So it was that majority (Republican) staff, as well as the Administration of President George W. Bush, opposed the inclusion of language mandating the use of participatory processes such as consensus conferences, but a Democratic representative worked to have this provision added in exchange for her support of the bill. This language adds considerable richness to the options that might be available for adding a significant reflexive capacity to NSE research institutions. Similarly, the language mandating the investigation of nanotechnology's implications for artificial intelligence reflected the concerns of a single representative. This language is important because it makes clear that social implications go beyond traditional risk-based formulations to broader considerations of desirability.

More generally, the social implications language in the 21st Century Nanotechnology Research and Development Act represents a response to the public debate that germinated around nanotechnology starting about the time when Bill Joy published his famous article. In some very real sense, Congress was acknowledging and seeking to address the nascent conflict and anxiety, in a way that perhaps reflected some learning from the bruising experiences of disputes over nuclear power generation, nuclear waste disposal, genetically modified foods, genomics, cloning.

Or perhaps not. We do not mean to be overly optimistic. As Fisher and Mahajan (in review) have noted, most of the 21st Century Nanotechnology Research and Development Act is devoted to accelerating the pace of NSE advance. The fact that the new law would allow—and perhaps even encourages—institutional innovation for democratic governance of nanotechnology does not suggest that such innovation will occur. In particular, it will be interesting to see if the current NSF competition for a Center for Nanotechnology in Society will lead to the testing of truly novel institutional arrangements, or if it will end up settling on a more conventional organization, perhaps with a more traditional risk-communication or public-understanding-of-science focus.

Whatever happens, it is abundantly clear that the community of scholars who have, over the past several decades, built up a deeply textured understanding of the social embeddedness of science and technology, were largely absent from the processes by which the social implications of nanotechnology became the increasing focus of federal attention and largesse. One cannot help but wonder if this lack of engagedness might have been a good thing, in that it allowed the "*common sense*" of Congressional staff to manifest in legislation that offers the potential for a truly innovative organization of social implications inquiry that potentially threatens the autonomy of NSE scientists and science studies scholars alike. Yet, as perversely satisfying as such speculation may be, it is perhaps closer to the mark to recognize this lack of engagedness as one of the reasons why nanotechnology got a 15 year head start on serious thinking about how society ought to govern its emerging capability in molecular manipulation. With Bill Joy's relinquishment on one side, and the NNI's full-steam-ahead approach on the other, there is plenty of room for creative experiments in scientific and technological governance rooted in theory and observation of scholars working in the fields of science studies and science policy. Two key questions remain: Are such scholars sufficiently willing to get their hands dirty? And is the momentum of nanotechnology still amenable to anticipatory governance? A "*yes*" to both may be necessary if we are to move beyond the brittle, reactive, regulatory governance modes that have characterized responses to technologies from nuclear power to genetically modified foods.

# 13

# ESSENCE OF NANOTECHNOLOGY

## EVOLVING INTO NANOTECH

If you take a look at the world around us, you will notice that nature herself designs at the molecular level. *Nanotechnology* intends to imitate nature by taking advantage of the unique properties of nanoscale matter to come up with more efficient ways of controlling and manipulating molecules. With technology, smaller is better. If you take a look at technical evolution, you will notice that we are continually getting smaller—computers the size of a room in the 1950s now fit on your lap; cellphones the size of a brick in the 1980s now fit in your shirt pocket. Consumer convenience, the economics of resources and competition, and the advantage of faster processing, higher productivity, and better quality all play a part in motivating companies to go small.

Technology is not only getting smaller, it's also evolving faster. As new technologies develop, we build upon previous knowledge. Thanks to the Internet, this knowledge base—and rate of information exchange—is increasing rapidly. To illustrate this evolution of technology, check out the evolution of electrical components: from the vacuum tube to the solid-state transistor to the carbon nanotube field effect transistor.

1. *Vaccum tube* (1897). This was the ancestor of the transistor, essentially a light bulb with three, instead of two, terminals. It was large, hot, and prone to burning out.
2. *Solid-state transistor* (1947). Instead of using a filament, the solid-state transistor switches between On and Off using different materials—metals and semiconductors. This let us come up with

transistors that were smaller in size, did not give off as much heat, and were far more durable.

3. *Carbon nanotube transistor* (1998). A carbon nanotube a graphite sheet rolled into a tube—comes in two main forms, metallic and semiconducting. The carbon nanotube was discovered in 1991 and within only seven years, was used for shuttling electrons across two electrodes. Not only is it incredibly small (nanoscale), but it also uses less energy and gives off less heat by using few electrons to indicate whether it's on or off.

It took 50 years to get from the vacuum tube to the first solid-state transistor—and another 50 years of refinement to get solid-state transistors to be all they could be. But when we'd developed the needed tools and understanding, it only took seven years from the discovery of the carbon nanotube to turn it into what may be the ultimate transistor.

*Nano,* Greek for "dwarf," means one billionth. Measurement at this level is in *nanometers* (abbreviated "nm")—billionths of a meter. To put this into perspective, a strand of human hair is roughly 75,000 nm across. On the flipside of the concept, you'd need ten hydrogen atoms lined up end-to-end to make up 1 nm. Figure 1-1 illustrates the differences in scale that range from you all the way down to one hydrogen atom.

**Definition**

*Nanotechnology* can be difficult to determine and define. For example, the realm of nanoscience is not new; chemists will tell you they've been doing nanoscience for hundreds of years. Stained-glass windows found in medieval churches contain different-size gold *nanoparticles* incorporated into the glass—the specific size of the particles creating orange, purple, red, or greenish colours. Einstein, as part of his doctoral dissertation, calculated the size of a sugar molecule as one nanometer. Loosely considered, both the medieval glass workers and Einstein were nanoscientists. What's new about current *nanoscience* is its aggressive focus on developing applied technology—and the emergence of the right tools for the job.

When faced with a squishy term that can mean different things to different people, the best thing to do is to form a committee and charge it with drawing up a working definition. In fact, a committee was formed (the National Nanotechnology Initiative) and the following defining features of nanotechnology were hammered out:

1. Nanotechnology involves research and technology development at the 1 nm-to-100 nm range.

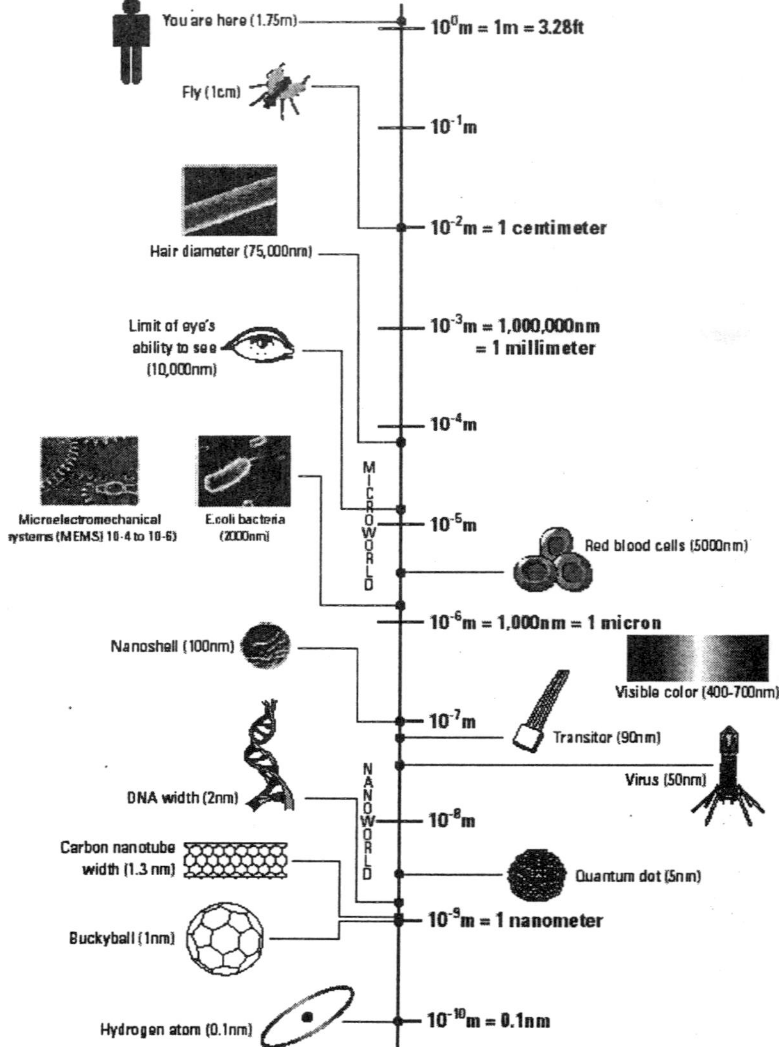


*Fig. 13.1. Size comparisons, from you (1.75 meters, or approximately 5 ft. 7 in.) all the way down to the hydrogen atom (0.1 nm).*

2. Nanotechnology creates and uses structures that have novel properties because of their small size.
3. Nanotechnology builds on the ability to control or manipulate at the atomic scale.

Numbers 1 and 3 are pretty straightforward, but Number 2 uses the eyebrowraising term "novel properties." When we go nano, the interactions and physics between atoms display exotic properties that they don't at larger scales. "How exotic?" you ask? Well, at this level atoms leave the realm of classical physical properties behind, and venture into the world of quantum mechanics. David Rotman described this best in his 1999 article, "Will the Real Nanotech Please Stand Up?"when he quoted Mark Reed, a nanoelectronics scientist at Yale University: "Physical intuition fails miserably in the nanoworld . . . you see all kinds of unusual effects." For example, even our everyday electrons act unusual at the nano level: "It's like throwing a tennis ball at a garage door and having the ball pop out the other side."

Need another concrete example? Check this one out. It is demonstrably true that a gold nanoparticle has a colour, melting point, and chemical property different from those you'd find in a macro-scale Fort Knox gold brick. That's because the interactions of the gold atoms in the larger gold brick average out—changing the overall properties and appearance of the object. A single gold nanoparticle, on the other hand, can be its own idiosyncratic self—a tiny object, free from the averaging effects of countless other gold atoms.

**Applications**

Nanotechnology is, at heart, interdisciplinary. You'll get only part of the story if you just use chemistry to get at the properties of atoms on the nano level—adding physics and quantum mechanics to the mix gives you a truer picture. Chemists, physicists, and medical doctors are working alongside engineers, biologists, and computer scientists to determine the applications, direction, and development of nanotechnology —in essence, nanotechnology is many disciplines building upon one another. Industries such as materials manufacturing, computer manufacturing, and healthcare will all contribute, meaning that all will benefit—both directly from nanotechnological advances, and indirectly from advances made by fellow players in the nano field. (Imagine, for example, quantum computers simulating the effectiveness of new nanobased medicines.)

There are two approaches to fabricating at the nano scale: top-down and bottom-up. A *top-down* approach is similar to a sculptor cutting away at a block of marble—we first work at a large scale and then cut away until we have our nano-scale product. The other approach is *bottom-up* manufacturing, which entails building our product one atom at a time. This can be time-consuming, so a so-called *self-assembly* process is employed—under specific conditions, the atoms and molecules spontaneously arrange themselves into the final product.

Some science-fiction plots—they know who they are—revolve around this self-assembly concept, conjuring up plot lines infested with tiny self-replicating machines running amok. For the near-term, it looks like the top-down approach will be favoured because it tends to provide us with greater control (and, more importantly, it uses some time-tested techniques of the computer industry). If we were betting men —which we are not, because as men of science we know that the House always wins—we would venture that the top-down approach will be the fabrication method of choice for quite awhile.

**History**

The word "nanotechnology" proper was coined by Nario Taniguchi in 1974 to describe machining with tolerances of less than a micron. But this really isn't the term's true beginning. Three noteworthy events and discoveries got this ball rolling—all by Nobel laureates (and when a Nobel Prize winner says something, you listen ... and try to understand).

1. *Vision*. In 1959, Caltech physicist Richard Feynman gives his famed talk "There's Plenty of Room at the Bottom," outlining the prospects for atomic engineering.
2. *Seeing is Believing*. In 1981, Gerd Binnig and Heinrich Rohrer of IBM's Zurich Research Laboratory create the scanning tunneling microscope, enabling researchers to both see and manipulate atoms for the first time.
3. *Nanostructures*. In 1985, Robert F. Curl Jr., Harold W. Kroto, and Richard E. Smalley discover buckminsterfullerenes, soccer-ball-shaped molecules made of carbon and measuring roughly 0.7 nm wide.

## Need of Nanotechnology

Nanotechnology will increase your standard of living—no ifs, ands, or buts. Done right, it will make our lives more secure, improve healthcare delivery, and optimize our use of limited resources. Pretty basic stuff, in other words. Mankind has spent millennia trying to fill these needs, because it has always known that these are the things it needs to ensure a future for itself. If nanotechnological applications pan out the way we think they will pan out, we are one step closer to ensuring that future.

## Security

Security is a broad field, covering everything from the security of our borders to the security of our infrastructure to the security of our computer networks. Here's our take on how nanotechnology will revolutionize the whole security field:

### *Superior, lightweight materials*

Imagine materials ten times stronger than steel at a fraction of the weight. With such materials, nanotechnology could revolutionize tanks, airframes, spacecraft, skyscrapers, bridges, and body armor, providing unprecedented protection. Composite nanomaterials may one day lead to shape-shifting wings instead of the mechanical flaps on current designs. Kevlar, the backbone fiber of bulletproof vests, will be replaced with materials that not only provide better protection but store energy and monitor the health status of our soldiers. A taste o what's to come: MIT was awarded a $50 million Army contract in 2002 to launch the Institute for Soldier Nanotechnologies (ISN developing artificial muscles, biowarfare sensors, and communications systems.

### *Advanced computing*

More powerful and smaller computers will encrypt our data and provide round-the-clock security. Quantum cryptography—cryptography that utilizes the unique properties of quantum mechanics—will provide unbreakable security for businesses, government, and military. These same quantum mechanics will be used to construct quantum computers capable of breaking current encryption techniques (a needed advantage in the war against terror). Additionally, quantum computers provide better simulations to predict natural disasters and pattern recognition to make *biometrics*—identification based on personal features such as face recognition—possible.

### *Increased situational awareness*

Chemical sensors based on nanotechnology will be incredibly sensitive—capable, in fact, of pinpointing a single molecule out of billions. These sensors will be cheap and disposable, forewarning us of airport-security breaches or anthrax-laced letters. These sensors will eventually take to the air on military unmanned aerial vehicles (UAVs), not only sensing chemicals but also providing incredible photo resolutions. These photos, condensed and on an energy-efficient, high resolution, wristwatch-sized display, will find their way to the soldier, providing incredible real-time situational awareness at the place needed most: the front lines.

***Powerful munitions***

Nanometals, nano-sized particles of metal such as nanoaluminum, are more chemically reactive because of their small size and greater surface area. Varying the size of these nanometals in munitions allows us to control the explosion, minimizing collateral damage. Incorporating nanometals into bombs and propellants increases the speed of released energy with fewer raw materials consumed—more (and better-directed) "bang" for your buck.

**Healthcare**

Making the world around us more secure is one thing, but how about making the world *inside* us more secure? With nanotechnology, what's beneath our skin is going to be more accessible to us than it's ever been before. Here's what we see happening:

***Diagnostics***

Hospitals will benefit greatly from nanotechnology with faster, cheaper diagnostic equipment. The lab-on-a-chip is waiting in the wings to analyze a patient's ailments in an instant, providing point-ofcare testing and drug application, thus taking out a lot of the diagnostic guesswork that has plagued healthcare up to now. New contrast agents will float through the bloodstream, lighting up problems such as tumours with incredible accuracy. Not only will nanotechnology make diagnostic tests better, but it will also make them more portable, providing timesensitive diagnostics out in the field on ambulances. Newborn children will have their DNA quickly mapped, pointing out future potential problems, allowing us to curtail disease before it takes hold.

***Novel drugs***

Nanotechnology will aid in the delivery of just the right amount of medicine to the exact spots of the body that need it most. Nanoshells, approximately 100 nm in diameter, will float through the body, attaching only to cancer cells. When excited by a laser beam, the nanoshells will give off heat—in effect, cooking the tumour and destroying it. Nanotechnology will create biocompatible joint replacements and artery stents that will last the life of the patient instead of having to be replaced every few years.

**Resources**

The only thing not in short supply these days is more human beings—and we're not about to see a shortage of them any time soon. If we are going to survive at all—much less thrive—we are going to need to find ways to use the riches of this world more efficiently. Here's how nanotechnology could help:

***Energy***

Nanotechnology is set to provide new methods to effectively utilize our current energy resources while also presenting new alternatives. Cars will have lighter and stronger engine blocks and frames and will use new additives making fuel more efficient. House lighting will use quantum dots—*nanocrystals* 5 nm across—in order to transform electricity into light instead of wasting away into heat. Solar cells will finally become cost effective and hydrogen fuel cells will get a boost from nanomaterials and nanocomposites. Our Holy Grail will be a reusable catalyst that quickly breaks down water in the presence of sunlight, making that long-wished-for hydrogen economy realistic. That catalyst, whatever it is, will be constructed with nanotechnology.

***Water***

Nanotechnology will provide efficient water purification techniques, allowing third-world countries access to clean water. When we satisfy our energy requirements, desalinization of water from our oceans will not only provide enough water to drink but also enough to water our crops.

## REVOLUTION

We predict that a few revolutions will roll over us in the course of the next 50 years—revolutions that will have great impact on our lifestyles and economy, and will involve nanotechnology, energy, and robotics. We've broken down each revolution here, complete with an estimated peak year for each one's public and financial popularity. Note, however, that nobody's jumped in a time machine to check up ahead; these peak years are based on the "*gut feelings*" of the authors, and on our research and observations of current trends. But we aren't just spinning tales. In February 2005, Business Week Online polled its readers, asking when they thought nanotechnology would change their lives: 34 percent said by 2007; an additional 51 percent said by 2015. At this point, it's not a matter of "if" but a matter of "when"—it's nearly as certain as death and taxes.

**Nanotechnology**

We're thinking 2012 is about the time that significant revolutionary products will be available, along with solid companies within the industry. A "nanotechnology bubble" will begin to develop around 2010, but this may not be as drastic as the "Internet bubble."Low-power, high-density computer memory, longerlasting batteries, and some medical applications (including cancer therapy and diagnostics) will be some of the early products. Advances in computer processing will follow (2015), and new materials and composites will come online toward 2020. The order of events will likely be computers and medical first, and then materials—all overlapping but peaking at their respective years.

**Energy**

Our crystal ball says 2025 is going to be the Nanotechnology Energy Year. With demand for energy rising in industrializing countries such as China and India, oil will continue to be that highly-sought-after resource. Oil prices and prosperity have an inverse relationship —as oil prices go up, prosperity goes down. Goldman Sachs has recently (April 2005) suggested that we may enter into a "*super-spike*" period of oil demand, with prices as high as $105 a barrel—almost twice the current price. It has also been suggested that world oil production will peak around Thanksgiving 2005. Unfortunately, production capacity has grown more slowly than demand, which makes things even worse.

Nanotechnology will combine efficient use of our current sources while providing directions to explore for alternate sources of energy. Nanomaterials that emerge around 2020 will not only provide lighter/stronger materials for vehicles but will also improve efficiency in the collection, storage, and transmission of energy, greatly aiding our transition from gas to solar, hydrogen, or maybe even renewable bio-fuels (for example, vegetable oils and bioalcohols such as ethanol and methanol).

**Robotics**

Think 2045. This may seem a little farfetched, because you expect something with the personality of C3-PO and the powers of a Jedi, but today you're getting R2-D2—just bells and whistles. However, there are a few driving forces making robotics economically feasible: defense, space exploration, and labour. In the near term, autonomous UAVs (short for Unmanned Aerial Vehicles) will keep continuous watch over our borders, and robots will dispose of roadside bombs in the battlefield. Space exploration will be done by robots, cutting the need for human involvement and thus allowing us to go farther than we've ever gone before.

As nanotechnology develops better sensors and processors and the energy revolution provides abundantly cheap energy, robots will be in demand as cheap manual labour, increasing our overall standard of living. Not only will we have robotic dogs and vacuum cleaners but also assembly- line industrial labour, bringing money back to Western nations. Perhaps a "*robotic arms race*" will emerge, not as a mighty military machine but as a productivity machine—each nation trying to make the cheapest goods as quickly as possible. All this will gradually grow over the next few decades—but once the hardware is in place (around 2030), the software and artificial intelligence will soon follow.

Some of these years may seem a long way off, but these changes will arrive faster than you may think. If you're currently in college, they'll happen in your lifetime.

## Expectation for Nanotechnology

Futuristic excitement aside, our expectations for nanotechnology need to be realistic and we need to be patient, for not all the advances that nanotechnology is set to bring will happen overnight. Nanotechnology will not be a miracle cure. Although there will be some fantastic advances, not everything that we imagine will come to fruition. However, nanotechnology is also sure to usher in things that we never envisioned coming—products that could end up changing the world.

*Nanoscale science* isn't a free-for-all—there are rules. We won't be able to manufacture something that, at the molecular level, is chemically unstable. Scientists know how most things work chemically and physically, but there have been a few surprises—and we learn the rules along the way. Time to take a look at some examples of the nanotechnology we now have, what we can improve upon, what will be new, and what (we can confidently say) will never happen.

### *What we have*

Nano applications are already showing up in areas as diverse as computing, transportation safety, and medicine. The steps may seem modest by future standards, but they get big effects from tiny things. Three examples illustrate what we can do now:

*Computer transistors* have broken below the 100 nm barrier—transistors are officially nano-sized. Look for the devices that house them to shrink as well, and for devices that are already small (such as cellphones) to become more powerful.

*Airbag sensors*, although micro in size and bigger than nano, are used in most recent cars—some of them already saving lives. These sensors will continue to shrink, becoming more powerful and accurate.

At least one *home pregnancy test* (Carter-Wallace's "First Response") uses both gold nanoparticles and micrometer-size latex particles on an external, disposable test sheet. The product takes advantage of how gold nanoparticles of different sizes reflect light differently. If a woman is pregnant, a specific hormone is present that causes the micro-sized and nano-sized particles to clump together—and those bigger particles reflect a distinctive colour: a visible pink strip appears on the test sheet. If she's not pregnant, no hormone is present, which means no clumping of the nanoparticles and no pink strip.

***What will be improved***

Besides introducing new products and procedures, nanotech will advance those that already exist. In January 2005, the Lemelson-MIT Program identified the top 25 innovations of the past 25 years. Nanotechnology was number 21 ... which is good. What's even better is that the other 24 would all be positively influenced by nanotechnology. Try these examples on for size:

1. Cellphones with longer battery life
2. Global Positioning Systems that are smaller and more accurate
3. Computers that are faster and smaller
4. Memory storage that packs greater capacity into a smaller space and uses less energy
5. DNA fingerprinting that is quick and accurate

Other items not on the list will be oil additives designed to get more out of our precious resource, new medical diagnostics and drug delivery ... even an aesthetically pleasing sunscreen. (Most sunscreens are a white, thick and sticky cream. Nanophase Technologies has developed a sunscreen that is transparent—the active ingredient is a nano-scale material that, because of its small size, doesn't scatter visible light.) *Nanotech* will crop up everywhere in existing products, even—or especially—in places you can't see.

***What will be new***

Nanotechnology promises to be a cornucopia of wonders—improving our healthcare, optimizing our use of resources, increasing our standard of living. Detecting disease at the molecular level will lead to new treatments for old ills. The development of materials ten times stronger than steel—but a tenth of the weight—offers to make transportation faster and more efficient. (Imagine, for example, how air transportation would change if airframes were lighter and stronger, plane engines used less fuel, and sensors and smart material automatically deformed the wings to minimize drag.) New, nanotechbased paints and coatings will prevent dirt and water from adhering to surfaces such as kitchen counters, vinyl siding, cars, and windows. (Imagine a car you never have to wash, that rolls dirt and water right off so you don't even have to use your windshield wipers.)

Speaking of cleanliness, EnviroSystems' EcoTru disinfectant cleaner is the only EPA-registered Tox Category IV disinfectant product—it doesn't harm the skin, eyes, lungs, or body if ingested. Conventional disinfectants dissolve in a solvent, and are meant to drown the organisms with toxic chemicals—which can be about as bad for humans as for small organisms. EcoTru uses nanospheres of charged oil droplets suspended in water to carry the active ingredient that ends up penetrating the microorganisms' membranes. This stuff is so good that Doctors Without Borders used EcoTru in the operating room as an antiseptic when they ran out of their regular antiseptic. Of the 500 patients that they used EcoTru on, none got an infection. EcoTru is also already used as a disinfectant on airplanes, on cruise ships, and in healthcare facilities, and, given Doctors Without Borders's experience, may be used as an antiseptic in the future.

***What will not happen***

Science fiction writers describe swarms of molecule-size robots swimming through your bloodstream cleaning your arteries while shooting cancer cells. And the nanobots that aren't fixing your body are out there fixing and building the world around us, one atom at a time. These scenarios are highly improbable, if not impossible. If they do eventually prove possible, they're decades (if not centuries) away.

As marvelous as it is to envision nanobots curing our bodies and quickly assembling and disassembling inanimate objects, these methods may not even be the most efficient approach. After all, some of the best medicine involves coaxing the body to help fix itself—and building inanimate objects one atom at a time (even something as simple as a chair) is no quick task. There may be a better, more inventive way to use engineering principles at the nano scale—one that takes advantage of the opportunities that chemistry and intermolecular interactions offer. But those opportunities are far more modest and (well, yeah) small-scale than science fiction would suggest.

The dramatic creation and transformation of macro-scale objects makes for spectacular entertainment but dicey science. Here's why: *Molecular chemistry* is very complex and involves controlling atoms in three dimensions. At each reaction site, the atoms feel the influence of neighbouring atoms. To do any mechanics at this level (which is what nanobots would have to do for this to work), you would need to control the motion of each and every atom—a very difficult juggling act.

A lot of this nano-zealous science fiction got started in 1986, when K. Eric Drexler, founder of the Foresight Institute (a nonprofit organization dedicated to educating the public about nanotechnology), penned *Engines of Creation: The Coming Era of Nanotechnology*. In it, he describes selfreplicating nanoassemblers building objects one atom at a time. He also describes a doomsday scenario referred to as "*gray goo*"—myriads of selfreplicating nanoassemblers making uncountable copies of themselves and consuming the earth. "*Gray*" because they're machines; "*goo*" because they're so small they'd look like a thick liquid. Scientists have since ridiculed this Drexlerian vision—even Drexler himself (in 2004) said "runaway replication" was unlikely.

In the December 1, 2003, edition of *Chemical & Engineering News*, Eric Drexler and Richard Smalley, Nobel Prize winner and discoverer of the buckyball, squared off, arguing for and against molecular manufacturing. The "Point- Counterpoint" article was a series of letters between the two, where Drexler continues to outline his "*mechanical*" molecular manufacturing, whereas Smalley argues against such a model by describing a need for "chemistry" even at the nano level. One source of contention is the "gray goo" scenario. Drexler had presented this scenario as a warning to not let nanotech get out of hand whereas Smalley sees it as unnecessarily scaring the public on a doomsday scenario that's a) highly unlikely and b) threatens public support for nanotech by harping the negative. In the end, they both disagreed on molecular manufacturing but continue to promote nanotechnology's potential.

## A Piece of Nanotechnology

*Nanotechnology* is set to insinuate its way into our economy in ways that we can only imagine—more probably, in ways that we could never dream. Nanotechnology as an industry will be hard to identify and track, considering it's very pervasive—it touches upon a lot of different industries. Familiar products will be the first to take advantage of nanotech—clothing, cosmetics, and novel industrial coatings. When the processes to make these products are mastered, new and innovative products will begin to emerge. Businesses will rise, and those that don't adapt to this changing environment will fall by the wayside. Now that you're salivating at the prospects of making a quick buck, eager to enter the Lilliputian world of atoms and molecules, this section of the chapter will temper your appetite and present what we believe to be a *realistic* picture of the current industry—and how it will probably play out.

**Nanotech Industry**

The National Science Foundation projects that nanotech will be a $1 trillion industry by 2015—that's 10 percent of the current GDP of the United States. Small wonder (so to speak) that the National Nanotechnology Initiative has increased federal funding for nanotechnology research and development from $464 million in 2001 to $982 million in 2005. Governments across the world poured $4 billion into nanotech research in 2004.

However, according to Lux Research Inc., only $13 billion worth of manufactured goods will incorporate nanotechnologies this year. In the grand scheme, that's not very much. Nanotechnology, because of its complexity and reach over different industries, will grow slowly over many decades. Therefore, we must be patient and not expect a huge explosion of products coming online all at once.

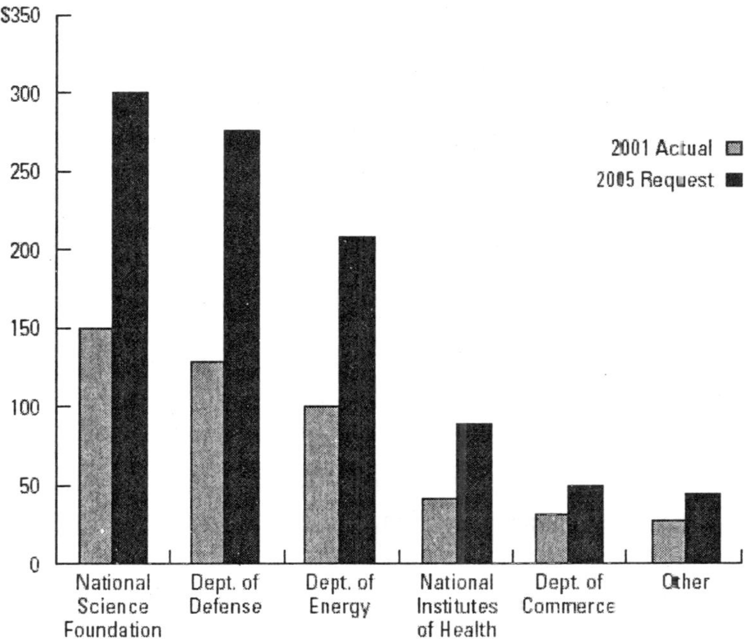


*Fig. 13.2. 2005 breakdown of nanotech expenditure among different government departments.*

Today, nanotech companies are either developing their knowledge base through research or are producing the materials for other people's nanotech research. There are three entities from which nanotechnology will emerge: open research (universities and national labs), large corporations, and startups. These entities, in turn, have three options: They can produce their discoveries, license them, or sell the rights to them outright. Licensing may be an attractive option–it creates cash flow with minimal overhead.

Big companies will have some advantage here, given their resources and ability to purchase high-end measurement equipment. Some large industries, such as pharmaceuticals and microchips, will be able to successfully integrate nanotechnology because they already have the processes and facilities in place to get their product to market—something small companies may not have access to.

However, this shouldn't discourage small companies who have the flexibility to adapt. Small companies are set to develop products, processes, and intellectual property, becoming attractive takeover candidates by big companies. In the end, everyone wins—investors make money, small companies develop products rapidly and efficiently, and big companies produce and distribute the end product to the consumer.

Both large and small companies, given nanotech's reach, will have to develop partnerships and collaborations—not only between different industries but also between different companies, universities, and government labs. Not only will they exchange research but also resources. For example, Rice University has a partnership with the Texas Medical Center, the largest medical center in the world. Additionally, Rice (as well as other universities across the nation) has broken down some financial barriers by developing the Shared Equipment Authority (SEA), which will train and allow businesses to use million-dollar measurement equipment for reasonable prices. If you want (for example) to learn how to use a scanning electron microscope, it will cost you $200 to get trained and $20/hour for each subsequent use. That's incredibly cheap, considering that the equipment costs $500,000 to begin with, and more than $40,000/year just to keep it maintained.

**Battle of the Bubbles**

Every industry has an *economic bubble*—speculation in a commodity that causes the price to increase, which produces more speculation, which causes the price to increase ... until, at some point, the price reaches an absurd level and the bubble bursts, causing a sudden—and precipitous—drop. Bubbles aren't good—they encourage people to misallocate resources in ways that don't pan out, with nonproductive results. Additionally, the crash that follows can cause great economic problems. Of course, if you don't get carried away, an economic bubble isn't all bad—it allows money to flow into a new industry to give it financial support to grow from. But too much blind enthusiasm too soon can backfire as market forces weed out ventures that don't produce actual profits in a reasonable time.

Many industries have had bubbles—the railroad industry, the automobile industry, even the tulip industry had an economic bubble. The most recent example was the Internet bubble in the late 1990s. Alan Greenspan, chairman of the Federal Reserve Board, identified this emerging bubble in 1996, calling investor speculation "*irrational exuberance*," which should have been enough to warn off most folks —but fools and their money are soon parted.

So, is it foolish to invest now in nanotechnology? Not necessarily. We have no doubt that there will be a "*nanotechnology bubble*", but we hope it will prove to be more "exuberant" and less "irrational." One thing to keep in mind is that, with the dot-com bubble, companies were started for $5,000 by lawyers and marketing agents who were pushing an idea — and not necessarily a product. Nobody was quite sure what to do with the idea yet. The essentials of the scientific knowledge base could be pretty well absorbed by your average information-technology technician after only a few months of training. Nanotechnology and nanoscience research, on the other hand, are a much more complex kettle of fish. They require in-depth scientific knowledge and a Ph.D.-level background. This knowledge base will be indispensable to any nanotech start-up company, ensuring that the field will not be inundated by every Tom, Dick, and Harry.

This requirement of high-level technical know-how is slowing nanotechnology's growth. There is a shortage of talent—a limited supply of scientists coming up against increasing demand. Additionally, product-cycle times—the time from research to market—is long. On the flipside, nanotech has fewer barriers to market adoption than the Internet had. In order for online stores to be successful, customers had to own a computer, establish an Internet connection, and gain confidence in online transactions. Nanotechnology will already be integrated into existing products—no massive new-productadoption process will be required.

Nanotechnology's biggest advantage over the Internet as The Next Big (Little?) Thing will be its patentable intellectual property. Web-based innovations were difficult to patent, allowing competitors to quickly adapt and clone a product within months. Nanotechnology's intellectual property creates a major barrier to entry by being incredibly difficult to replicate. The time cost to replicate is measured in years, encouraging the competitor to either take a different approach or license the patent. This barrier to entry is fantastic for small companies; they have more time to develop without getting immediately crushed by the big dogs.

**Caveat Emptor**

If recent history is any indication, our coming nanotech bubble may be overhyped and fueled with stock speculation. In this heading, we hope to quell some of this speculation early on and paint a realistic picture for you. Currently, very few nanotech companies are public —and the ones that are have a stock chart that reads more like a cardiogram than a steady line of growth. That hasn't deterred some early players; on the Dow Jones Industrial Index, 19 of the 30 companies have launched nano initiatives. This heading takes a look at two particular companies and throws in some indexes you can follow as well.

***Nanosys***

In April 2004, Nanosys filed for a $115 million initial public offering (IPO) with the Securities and Exchange Commission (SEC). They are a company rich in patents—over 200—but alas, no profits yet (sigh). This is not to say they won't ever make money—they are one of the companies that our research shows has some great potential —and it's a pretty sure bet they'll license their research.

For now, they're cautious. In the *Nanosys SEC filing*, the company stated: "To date, we have not successfully developed any commercially available products. ... We do not anticipate that our first products will be commercially available for at least several years, if at all." (*If at all?* Not the rosiest picture.) Nanosys even withdrew its filing in August 2004, stating "*volatility of capital markets*." But *Nanosys* has clout as one of the poster-child nanotech companies—and we give them credit for not starting a wild nanotech bubble before the industry has had time to mature.

***Altair Nanotechnologies, Inc.***

On February 10, 2005, *Altair Nanotechnologies*, Inc., announced a breakthrough in their lithium-ion battery-electrode material. This novel nanomaterial—composed of nano-size lithium-titanium-oxide particles —does the following for their rechargeable batteries:

1. *Delivers more power.* You get three times the power of existing lithiumion batteries, to be exact.
2. *Allows faster recharge.* Recharge takes a few minutes instead of hours.
3. *Ensures longer life.* The material improves the number of recharge and discharge cycles from a few hundred to many thousand cycles.

On this news, the stock spiked. Here's what the results looked like:

1. *Day before.* Closed at 2.08 with volume of 500k.
2. *Day of.* Closed at 4.77 (129.3% increase) with volume of 56.6 million.
3. *Day after.* Peaked at 6.52 (another 36.6% increase) with volume of 101.6 million.

Not bad for two days' worth of work. As of mid-April, 2005, ALTI had a 50-day moving average of 3.8 with an average volume close to 10 million. On April 19, 2005, they announced the initial shipment of their electrode nanomaterial for testing at a partner company, Advanced Battery Technologies, Inc. This nanomaterial will be incorporated into Advanced Battery's Polymer-Lithium-Ion (PLI) batteries tested in electric vehicles.

***Investment tools and strategies***

Whether or not Altair Nanotechnologies or Nanosys are going to be the next eBay is certainly up in the air. Given the volatility of owning individual stocks in nanotechnology, it's wise to go with something a little more diverse, such as an exchange-traded fund (ETF) or mutual fund. Unfortunately, there are no nanotech ETFs or mutual funds at this time. Harris & Harris is as close as you're going to get for now; it's a publicly traded venture-capital firm specializing in nanotech companies.

Merrill Lynch established a "*Nanotech Index*" in 2004 — on April Fools Day, of all days (those guys are such kidders). The index consists of 22 small companies whose future business strategy is based on nanotechnology. Lux Research, Inc., is a research and advisory firm focusing on the economic impact of nanotech businesses. Their index, Lux Nanotech Index, follows 26 publicly traded nanotech companies. Merrill Lynch's Nanotech Index is open for trading, but Lux Nanotech Index is not.

As you conquer your greed and quell your fears (both wise moves), here are a few generic tips to keep in mind.

1. *Follow the money*. There will undoubtedly be huge investments into nanotech. Be mindful of large volume shifts, indicating large institution investment.
2. *Follow the trend*. 75 percent of all stocks follow the Dow Jones Industrial Average—this will include nanotech stocks. If stocks are going up, buy; if going down, sell. Buy low, sell high.
3. *Don't listen to tipsters*. Tipsters will, more than likely, be wrong about a lot of nanotech. We still need those scientists to help business analysts make sense of what's plausible and realistic. Do your own research; don't let a news sound bite influence your decision and control your money.
4. *Set a stop loss*. Set a sell-point with your brokerage firm; make it roughly 7 percent below your purchase price—and as the price goes up, increase this stop-loss point. Doing so prevents what happened to many when they rode the 1990s Internet crash into the ground.

# 14

# NANO AWARENESS

From the development of the earliest stone tools to the must sophisticated *microprocessor*, man has increasingly, and sometimes unwittingly, shaped the world around him through the use of his technologies. These technologies impact upon all aspects of our lives. We depend upon them for the food we eat, our transport and our communications. We rely on them for clean water and an increasingly sophisticated level of health care. Whole periods of human history are labelled by reference to the dominant technology of the time—the stone age, the bronze age, the iron age, the industrial age, the computer age. We are all familiar with these terms and use them without thinking about die profound effect that each of the technologies had—both upon the societies that created them and on the planet itself.

Frequently, we are not aware of the impacts the older technologies have had on the world in which we now live—for example, stone axes were used to fell the ancient forests that once covered the UK. and created the downs and pasture that we now recognize as our "green and pleasant land". This chapter will look at one new area of technology—*nanotechnology*—and attempt to answer the question "Should we be worried—either about what it is doing now or where it is taking us?" I wonder whether a Neolithic farmer stopped to ask himself the same question while he was felling the trees to create a new stretch of farm land for himself and his family.

Before discussing the issue of where nanotechnology is taking us, and whether we should be worried about it, we must try to understand what it is. So, the first question that is to be addressed is: "What is nanotechnology?" This is not as easy to answer as one might think, because the term encompasses a huge range of activities. Some people think it is not a single type of activity at all, while others think it is just a term that 'has been invented to allow researchers to extract In amounts of research funds from government funding agencies! Nanotechnology has received enormous attention in the last 15 years, and especially so recently; even the heir to the British throne has become involved.

Some commentators and financial observers—such the finance house Merryl Lynch—have even gone so far as to suggest that the impact of nanotechnology will be so great that the term will be used to describe a new era of world economic growth. Why all the fuss? What is this phenomenon that everyone is getting so excited about? People are far too concerned about these—particularly so with nanotechnology. The prefix "*nano*" comes from the Greek word "*nanos*" meaning "*a dwarf*. Hence "*nanotechnology*" might well simply mean a technology concerned with small things. However, "'nano" has also long been used as a prefix in scientific circles to mean one billionth (using billion in its American sense of a one followed by nine zeros). So we have the term "*nanogram*" for one billionth of a gram and nanometer for one billionth of a meter.

A nanometer is exceedingly small—only about 10 atoms across. On that score, we might expect "*nanotechnology*" to have something to do with technologies that are working, for example, at the nanometer level and this is the general sense in which the term nanotechnology is used today. It is important to distinguish here between nanoscience, which is the study of phenomena at the very small scale, and nanotechnology, which implies an aim to achieve an end that is in some way "*useful*". The Royal Society/Royal Academy of Engineering Working Group on the subject adopted the following definitions:

*Nanoscience* is the study of phenomena and manipulation of materials at atomic, molecular and macromolecular scales, where properties differ significantly from those at larger scale.

*Nanotechnologies* are the design, characterization, production and application of structures, devices and systems by controlling shape and size at nanometer scale.

The scale of sizes normally discussed for the applicability of nanotechnology is usually less than 100 nm. Nanoscience, arguably, has been around since the early part of the 20th century, while the idea that there might be some technological advantages to be gained by working at the very small scale came much later. It was first put forward by the famous physicist Richard Feynman, when he gave a lecture in 1959 to the American Physical Society entitled "There's plenty of room at the bottom—an invitation to enter a new field of physics". In this lecture—which actually had almost no physics in it but was mainly concerned with the technology of making things—he explored the benefits that might accrue to us if we started manufacturing things on the very small scale. The ideas he put forward were remarkably prescient. For example, he foresaw the techniques that could be used to make large scale integrated circuits and the revolutionary effects that the use of these circuits would have upon computing. He talked about making machines for sequencing genes by reading DNA molecules. He foresaw the use of electron microscopes for writing massive amounts of information in very small areas. He also talked about using mechanical machines to make smaller machines with increasing precision.

Interestingly, for a man who went on to win the Nobel Prize in Physics in 1965 for his work in the field of quantum electrodynamics, he hardly mentioned the quantum mechanical effects associated with the making of very small things, although he did talk about using the interactions of quantized spins, a kind of '*spin logic*', which is only now being studied. Many of his predictions in that lecture have come true, and all are aspects of what we would now call "*nanotechnology*", although he did not use the term itself. Actually, the first use of the term "*nanotechnology*" was by Norio Taniguchi who, in 1974, gave a talk describing how the dimensional accuracy with which we make things has improved over time. He studied the developments in machining techniques over the period from 1940 until the early 1970s and predicted (correctly as it turned out) that by the late 1980s techniques would have evolved to a degree that dimensional accuracies of better than 100 nm would be achievable. He called this "*nanotechnology*".

Incidentally, Cranfield Precision Engineering was one of the leading companies in helping to develop machines that could actually make things to this kind of precision. You will realize from the above that all the early running in the field of nanotechnology was made by physicists and engineers who mainly thought in terms of making things more and more precisely. This means using one machine to make another, usually smaller machine to greater precision, and using machines (frequently very large and expensive machines) to make things which have incredibly precise features defined upon them. We now call this "*top-down*" nanotechnology. It has led directly to the hugely successful semiconductor and information and communications technology ((ICT) industries, with a world market size in excess of $4 $\times$ $10^2$, currently growing at 4.8% per annum. We all use advanced microprocessors in our portable computers. These have metal lines written on them that are only 90 nanometers wide and have upwards of 100 million transistors on a single piece of silicon a few millimeters across. They are objects of mind-boggling complexity and yet they are manufactured at incredibly low cost—a few tens of dollars each.

The technologies established by the semiconductor industry are also now being applied in the manufacture of tiny micromechanical machines for sensing and actutation. These "*microelectromechanical systems*" (also known as MEMS) are finding their way into a host of applications, particularly in the automotive and medical fields, where cost and size-based functionality are key factors. We can start to see here the enormous effects that working at the very small scale is having on our world.

There were two other themes that Feynman put forward in his lecture. He envisaged the possibility of making machines that could pick up and put down single atoms. Putting particular atoms together in particular combinations would be a new way for making chemical compounds. Feynman did put forward several reasons why atomic manipulation might not work. These included the van der Waals and chemical forces that would make an atom stick to the finger picking it up, so that once picked up it would be virtually impossible to put down, let alone put it in a particular place relative to another atom. This has been called the "*sticky fingers problem*".

In 1981 Binnig and Rohrer, based at IBM in Zurich, invented the *scanning probe microscope*. This uses a very sharp metal point scanned over a surface to create images of the atoms in the surface. Incidentally they won the Nobel prize for this work in 1986. In 1989 Don Eigler used the scanning probe microscope to nudge atoms of xenon on a copper surface held at a temperature close to absolute zero to spell out the letters "IBM". Another one of Feynman's predictions had come true, admittedly under very special conditions. Eigler and his group have since done some remarkable work, mainly using the technique to explore basic physical and quantum mechanical phenomena. Jim Gimzewski at IBM used similar techniques to push single molecules around on surfaces. This kind of work with single atoms and molecules is called "*extreme nanotechnology*".

Feynman had another vision in 1959, which was of a factory in which billions of very small machine tools were drilling and stamping myriads of tiny mechanical parts, which would then be assembled into larger products. In the late 1980s another worker in California, Erie Drexler, combined these small manufacturing ideas of Feynman with another thought experiment, which had been put forward by John von Neumann in the late 1940s. This was the idea of a mechanical machine—called a "*clanking replicator*"—that could be programmed to make replicas of itself. All it would need was a supply of raw materials and a source of energy. Those replicas would make more replicas, and the result would be exponential growth in the number of the machines—until either the source of raw materials or the energy was exhausted.

Drexler combined this "clanking replicator" idea with Feynman's to come up with the concept of the "*universal assembler*", first put forward in his book "*Engines of Creation*". The clanking replicator concept is reduced to very small size through the use of mechanical components that are made on the molecular scale. The first assembler would be programmed to make copies of itself by atomic and molecular manipulation. The exponential growth would lead to billions of assemblers that would be programmed to work in concert with each other to build virtually anything required. You could sprinkle a few assemblers onto a heap of garbage and out would come a washing machine! Clearly such a technology would have enormous economic implications, both in terms of material and energy use, effects on employment, etc. Drexler has set up the Foresight Institute in California and has raised large amounts of money for nanotechnology research based on that idea. He also was the first to raise alarm bells about the possibility of the assemblers reproducing out of control and reducing everything to a "*grey goo*" of tiny machines.

There is now a new variant on this vision, which postulates the fusion of nanotechnology with biotechnology to create an assembler that is at least partly biologically based. In this case, the problem becomes one of "*green goo*" rather than "*grey goo*", but the outcome is essentially the same. This idea was subsequently hit upon by the author and screenplay writer Michael Crichton who used it in his novel "Prey", in which a set of assemblers, originally created for medical inspection and subsequently military purposes, goes out of control and starts hunting down and destroying their creators. There have been some remarkable developments in materials science and chemistry over the last 15 years or so, particularly where small size plays a big role in determining basic properties. In the field of materials science, size does indeed matter! If we take a piece of a semiconductor less than about 100 nanometers across, then the electrons in them behave differently from in the bulk. For example, the colours of light absorption and emission change.

Very small particles (*nanoparticles*) of materials like cadmium telluride are being used in applications such as the labelling of biological molecules and in new types of displays. These can be made amazingly precisely in size— say 50 nanometers, plus or minus a couple of nanometers—using reasonably standard wet chemical processes. Very small particles (less than a few hundred nanometers in size) do not scatter visible light. Good absorbers of ultraviolet light such as titanium dioxide are now being made in nanoparticulate form for sunscreens. The fact that the particles are so small means that they are invisible on the skin, while still being highly effective as UV blockers.

Very small particles also possess high surface areas per unit of mass. Oxonica, a start-up company from Oxford University, has found that nanoparticles of cerium oxide, when introduced into diesel fuel, act as oxidation catalysts during combustion. This provides improvements in fuel efficiency of up to 10% and reduces the emissions of carbon soot from the engine exhaust. If we look at other areas of materials science, we see that new forms of carbon have been discovered. Harry Kroto from the University of Sussex, together with Richard Smalley and Robert Curl, discovered the $\text{carbon}_{60}$ molecule in 1985 (and won the Nobel Prize for chemistry in 1996). It is a sphere 0.7 nanometers across that looks like a soccer ball, or the geodesic dome structure pioneered by the 1930s architect Buckminster Fuller, so they called it *buckminsterfullerene*.

It is amusing to note that if you could expand the $C_{60}$ molecule so that it was the same size as a soccer ball, the soccer ball itself, if blown up by the same factor, would be about half the size of the planet Jupiter! The so-called *fullerenes* form a whole family of related structures that possess remarkable physical and chemical properties. If fully fluorinated, the molecules, which can then be thought of as tiny Teflon balls, form one of the best lubricants known. In 1991 Iijima discovered carbon nanotubes. These are like sheets of graphite roiled into long tubes, each one being terminated by a fullerene group. They also have remarkable properties. They can be either metallic or semiconducting, depending on the precise way in which the carbon atoms are assembled in the tube.

The metallic forms have electrical conductivities 1000 times better than that of copper and are now being mixed with polymers to make conducting composite materials for applications such as electromagnetic shielding in mobile telephones and static electricity reduction in cars. They possess mechanical properties that are many times superior to those of steel, bringing the promise of replacing carbon fibres in a whole new generation of high 'strength composite materials. They have been demonstrated in applications as diverse as supercapacitors for energy storage, field emission devices for flat panel displays and nanometer-sized transistors.

Clearly, these nanomaterials hold huge promise for the future. So, what is nanotechnology? Is it a real subject, or is it just, as the cynics would .have us believe, a mechanism for extracting funding from gullible government agencies and investors. It employs all the conventional scientific and engineering subjects in order to achieve new applications. It does this through the exploitation of phenomena in which small size is the key to obtaining an exploitable property. Secondly, it is an area of endeavour where there are real and remarkable properties that we can seek to exploit. Thirdly, and increasingly, we are starting to see convergence between different areas of nanotechnology. The size range of interest between a few nanometers and 100 nanometers is one where many interesting things happen. All sorts of physical properties change and many biological systems function at this length scale. Hence, we are starting to see the use of processes such as electron beam lithography, originally developed for writing very fine scale features on silicon for electronics, being applied for the modification of surfaces on which biological species can be grown in a controlled way.

The self-assembling properties of biological systems, such as DNA molecules, can be used to control the organization of objects such as carbon nanotubes, which may ultimately lead to the ability to grow parts of an integrated circuit, rather than having to rely upon expensive top-down techniques. This type of self assembly is called "*bottom-up*" nanotechnology. It is hoped that the engineering complexity of integrated circuits means that the *top-down methods* will be with us for a long time to come but self-assembly techniques may have an increasing part to play. We are thus seeing an area that is providing real potential in its applications. Some people have predicted a world market for

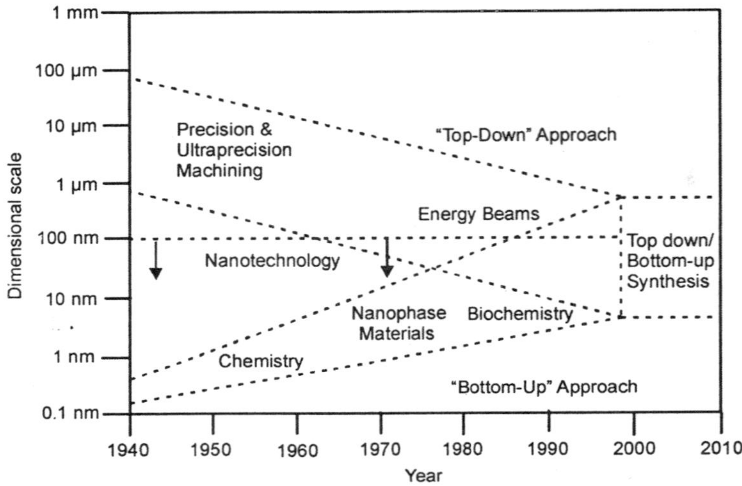


*Fig. 14.1. Illustrating the convergence of top-down and bottom-up nanotechnologies.*

nanotechnology-related products of billions or trillions of dollars by the end of the decade. There is no doubt that the technology associated with our ability to manipulate matter on the very small scale is already having major impacts on our lives and this impact will only increase. Hence, we should now ask the question that our *Neolithic farmer* (presumably) didn't: "Should we be worried?" Could the introduction of nanotechnology have unforeseen consequences? First let's consider the *Drexlerian dystopia* in which a rogue molecular assembler, ostensibly created for the betterment of mankind via high efficiency, low cost manufacture, goes out of control and reduces everything to a *grey* (or *green*) *goo*.

There are many highly rated, first class scientific minds (including Richard Smalley, the Nobel Prize winner referred to earlier) who have asserted that the "assembler" is not possible for all .sorts of reasons. These include the "*sticky fingers*" problem, problems with the storage and transmission of the huge amount of information needed and the vast complexity of the problem, which would be far, far greater than the complexity of a modem microprocessor. They imply that the assembler concept needs to stay where it belongs, firmly in the realms of Michael Crichton's science fiction novel. There are already very real self-assemblers all around us that owe nothing to nanotechnology. They are called *viruses* and *bacteria*. They pose a very real threat, that is moreover increasing, due to our farming practices, profligate use of antibiotics and cheap international travel. The creation of this hazard obviously cannot be laid at the door of nanotechnology, although some of the nanoscale analysis techniques evolved for nanotechnology may well be able to help with combating it.

What about the less spectacular aspects of nanotechnology? Are there aspects of these materials that we need to be careful about? There is no doubt that the properties that give nanomaterials their technological exploitability might also give us cause for concern. We have already seen in the last 100 years how a single material with hugely beneficial properties can bring equally huge problems. Asbestos was very widely used in the period between the late 19th and mid-20th centuries. It has extraordinarily useful insulating properties. The high efficiency steam engines used in ocean-going liners and steam locomotives would have been impossible without it. However, we all now know how lethal a few asbestos fibres can be if inhaled, causing occupational cancers, and many people have died prematurely because of it. Is there any risk that any of the materials that are emerging from our nanotechnology labs might be building-up a similar problem for the future? There are certainly enough physical similarities between the dimensional characteristics of asbestos fibres and carbon nanotubes to cause some concern. At the moment we just don't know whether carbon nanotubes are hazardous; there has not been sufficient research. There has been a small amount of work in which rats inhaled carbon nanotubes, but the results were inconclusive.

It is important to note that in all of the applications such as in composites or displays, the tubes would be very firmly fixed in a stable structure and would therefore be unlikely to pose a threat to the general public, although we should still ask questions about how the product would ultimately be disposed of. Would there be a chance of carbon nanotubes being released into the environment? We should also question how the products would be manufactured and what the exposure might be for people involved in the manufacturing (or, indeed, for researchers using the materials). A full life cycle analysis is essential for these new materials.

What about nanoparticles? Should we be concerned about inhaling these, or indeed rubbing them onto our skins? There is good evidence that we have been exposed to certain types of *nanoparticles* in the atmosphere for millennia. Wood-based camp fires are excellent sources of *nanoparticulate soot*, for example. There is little cause for alarm in the sense that nanoparticles, *per se*, do not constitute a new hazard at low levels of exposure. However, there is good evidence that heavy exposure to carbon black is a serious industrial hazard, and that heavy exposure to nanoparticulate soot from sources such as diesel exhausts may be a cause of cancer. There is also evidence that, should nanoparticles arrive in the lungs, they do not remain there, but readily cross the barrier into the blood stream, whence they can migrate to various parts of the body, including the brain. There is, unfortunately, very little evidence to tell us what damage nanoparticles might cause there.

On the specific issue of sunscreens, it is known that the main oxides of interest are inert when present at the larger scale, but again there is a dearth of evidence about how the particles might behave when at the nanoscale. There is as yet, no evidence lo suggest that they can penetrate into healthy skin, but the fact that nanoparticles are known to have .different properties from those exhibited in the bulk must give us some cause for concern should, they be ingested. There is a need for much more research in this area. We should consider treating size as a property when dealing with chemicals in, for example, assessments of chemical or disposal hazards, just as we would deal with any oilier chemical property. There is also a need for companies to put the results of research on nanoparticle-based products into the public domain. The uncontrolled environmental release of nanoparticles is another cause for concern. Nanoparticles that are not bound to other materials may become widely dispersed in the environment and turn up in unexpected places.

However, we must also keep a sense of proportion. We have little evidence at the moment that oxide nanoparticles form a general hazard and there may be many benefits in using them. If a nanoparticulate oxide fuel additive can reduce nanoparticulate carbon emissions, then that may have significant benefits for the health of the general population. An improvement in fuel efficiency would make a positive contribution through reductions in carbon dioxide emissions and thereby a reduction in global warming. Similarly, if the availability of nanoparticulate sunscreens were to increase their general acceptability and thereby their use, then this might contribute to a reduction in skin cancer. What is certain is that we need to understand more about the health and environmental issues associated with nanoparticulate materials.

Are there other areas for concern about nanotechnology? *Nanotechnology*, like virtually all technologies, is being considered for the contributions it can make to defence. These include obvious developments such as improved electronic systems for communications, improved sensors and, especially, improved materials. Examples include composites using *carbon nanotubes* and a novel, body armour that uses *oxide nanoparticles* dispersed in a fluid and held between two flexible *Kevlar sheets*. The composite fabric is flexible, but any attempt to penetrate the sheets by, for example, a projectile or other weapon causes the fluid to rapidly set into a rigid mass that protects the wearer. The applications to flexible, wearable, protection that would protect a soldier's limbs as well as his torso are obvious. *Metallic nanopowders* can be expected to deliver more powerful conventional explosives. These developments would certainly give the armies using them a military edge, but hardly constitute the kind of doomsday scenario that we sometimes read about in the popular press or see on our video screens in movies like Terminator II. Those images owe more to science fiction based on the *Drexler vision* than a level-headed scientific analysis of the advances that nanotechnology is really likely to deliver.

Paradoxically, the broad contribution that nanotechnology can potentially make to economic development has sometimes been cited as a cause for concern, especially by some pressure groups. They point out that the promise of nanotechnology would widen the disparity in development between the rich and poor nations of the world. The rich will get richer, while the poor will be unable to benefit from the *nanotechnology revolution*. Another criticism that has been raised against the nanotechnologists is that they are promoting their research as possibly providing cures for disabilities such as deafness. Advanced cochlear implants based, on MEMS technologies can already provide profoundly deaf people with the ability to hear.' Such implants are much more effective if they are given to children rather than to adults because the child can more quickly adapt to the inputs from the implant and learn how to use it. However, the challenge issued by some deaf people is to say, "I am perfectly happy with the way I am. What gives you, the scientist, engineer or doctor, the right to say that I—or more particularly my child—-ought to be "cured" of something that I do not consider as a disability?" Is it society that needs to change its attitude to those who are different?

These questions of how we as a society choose to deploy or employ the capabilities that nanotechnology may provide are very good ones. But it strikes me that they are ethical issues that pertain more to the general capabilities that our modem technologies can deliver, rather than being specific issues concerning nanotechnology *per se*. They are, perhaps, more in the realm of the philosopher than the engineer. Nevertheless, the success of the nanotechnologists in promoting their area makes them natural targets for these ethical questions. We nanotechnologists need to make links with the people who can help us to answer them, and this is starting to happen.

One clear area of concern for the nanotechnologist is the perception of the subject by the general public. We certainly want to avoid the problems that have beset the *genetically modified* (GM) area, largely through the attempt by large companies to impose *GM crops* and *foodstuffs* on the general public, without first entering into a dialogue with civil society representatives or attempting to engage people in debate. The approach backfired badly; especially in Europe. A recent survey conducted early in 2004 under the auspices of the Royal Society and the Royal Academy of Engineering showed that few people had any preconceived ideas about nanotechnology. Those who had heard of the topic were more- or-less equally divided between those who generally thought of it in terms of a beneficial high technology (better computers and mobile phones) and those who vaguely perceived it as slightly threatening or menacing. We are certainly not (yet) in the *GM position*. This is not to say we should be complacent. Nor is it correct to say that "if only the general public understood more about our subject, they would warmly embrace it." This so-called "*deficit model*" of the public understanding of science is rather controversial. Nevertheless, scientists and engineers certainly need to work harder to get the public, and especially our children, more engaged with the subject. This is a real area for concern, but again it is a general one and not simply to do with nanotechnology alone.

In this chapter efforts have been made to show how nanotechnology—the exploitation of matter when it is deliberately structured at the very small scale—has the potential to provide huge benefits, just as any useful technology should. It is a real, very broadly based and multidisciplinary area of human endeavour and not just a token epithet that can be applied to the latest research proposal or business venture in an attempt to get it funded, although admittedly it is frequently used that way. Certainly, there are issues that should concern us. These can be specific. We need, for example, to understand more about the health and environmental impacts of our uses of nanoparticles. There are also more general concerns. Nanotechnology could provide us with a broad range of capabilities but they need to be applied in a thoughtful and responsible manner. However, these concerns are similar to those that might have been applied to any significant technology in the past. The only difference is that we are now in a position to learn from history and can try to take action before mistakes are made. It is generalized suggestion that the action should be both proportionate and based on a realistic analysis of the likely risks and benefits of nanotechnology.

# 15

# SUBSTRATES FOR NANO

The union of distinct scientific disciplines is revealing the leading edge of Nanotechnology. Fifteen to twenty years ago, the inter-disciplinary activity of geneticists, biologists, immunologists and organic chemists created a more diverse toolbox now known as life science. *Bioconjugates* were created to help us move from outside the cell to the inside. Enabling technologies brought about the ability to create, identify and specifically manipulate genetic maps to engineer designer proteins. In parallel, physicists, chemists, polymer chemists and engineers were creating the foundation for the small world of nano materials science. *Fullerenes*, *carbon nanotubes* and *atomic force microscopy* were in their infancy.

In less than a decade, materials science and life science together are unraveling the mysteries of controlling, on a molecular level, the structure of matter. Particles, complexes, tubes, coatings, active surfaces and devices are being explored on the nanoscale. Assembly of nature's building blocks (e.g. carbon, nucleic acids, lipids and peptides) along with the combination of different materials (e.g. CdSe/ZnS, Au, Ag, Si(n)OH(n), light harvesting dendrimers and thin films) are leading to insightful understanding and the creation of new scientific tools. Chemists and physicists have been manipulating matter on the molecular level for centuries. Some say this is nothing new.

When one looks at the absolute elegance of the nanometer scale biological system, however, one is compelled to create, understand, manipulate and control systems with equal elegance. This high level of activity and promise has attracted private and government funding with considerable economic impact and growth. Since the launch of the National Nanotechnology Initiative in 2000, there have been hundreds of start-up companies emerging in the market. At universities, there are increased investments in nanotechnology programs and facilities. Northwestern University's Institute for Nanotechnology is a direct result of the high level of activity and pioneering work of Prof. Chad Mirkin.

As a result of massive grant support and continued focus from the university, the Center for Nanofabrication and Molecular Self Assembly building has been constructed. Similarly, The Molecular Foundry, a Department of Energy Nanoscale Science Research Center, is under construction at Berkeley in California. The research center is focused on the dissemination of nanoscale techniques and methods to enable scientists to delve into nano research. The Molecular Foundry is the direct result of Prof. Paul Alivasatos' discoveries and benchmark work. We see an increase in programs, facilities, career opportunities and educational outreach. It is a massive explosion of activity in study on a very small scale. Science is beginning to set free nanotechnology. Soon the only limit will be the imagination.

## Nanoparticle-catalyzed Growth of Semiconductor Nanowire

### Answers from the Past

Over the past twenty years, homogeneous nucleation in solution has proven to be an effective method for the synthesis of both *metallic* and *semiconductor nanocrystals*. Nearly *isotropic*, *pseudospherical* morphologies are typically produced. Other nanocrystalline morphologies are often desirable, and homogeneous nucleation has been adapted for the synthesis of rod-shaped morphologies. However, the growth of nanowires by homogeneous nucleation is apparently limited to materials having particularly favourable, highly anisotropic crystal structures. How then might nanowire morphologies be generally prepared?

The answer was uncovered by Wagner and Ellis in 1964, who discovered that micrometer-scale silicon whiskers (wires) could be grown from gold-droplet catalysts under *Chemical Vapour Deposition* (CVD) conditions at about 1000 °C. The process was named the "Vapour-Liquid-Solid" (VLS) mechanism after the three phases involved. VLS growth was extended to micrometer-scale whiskers of many inorganic materials and intensively studied for over a decade, before fading into relative obscurity.

### From VLS to SLS

In 1995, my group reported that III-V semiconductor nanowires could be grown in solution from indium nanoparticle catalysts by a process analogous to the VLS mechanism, and by analogy we named it the "*Solution- Liquid-Solid*" (SLS) mechanism. Shortly thereafter, Lieber and coworkers described laser-ablation adaptations of the VLS mechanism that afforded silicon and a wide range of other semiconductor nanowires (*Laser-Assisted Catalyzed Growth*, LCG). Many others made seminal contributions to nanowire growth by VLS adaptations, most notably Korgel (*Supercritical Fluid-Liquid- Solid, SFLS, mechanism*) and Yang. We now routinely prepare soluble, III-V and II-VI semiconductor nanowires with controlled diameters in the strong-confinement regime of about 3–20 nm by bismuth-nanoparticle-catalyzed SLS growth. The nanoparticle-catalyzed VLS mechanism and its solution-phase variants have emerged as the popular, widely practiced, general methods for the synthesis of semiconductor nanowires. Gold nanoparticles are presently by far the most commonly employed catalysts.

**Quantum Wires to Nanophotonics**

The advantages of nanoparticle-catalyzed nanowire growth include its general applicability to a wide variety of materials, the diameter control afforded, the uniformity of the wires (lack of significant diameter fluctuations) and their oriented, near-single crystallinity. The surface passivation, solubility and length of the wires may also be systematically varied. Small-diameter "*quantum*" wires are ideal specimens for fundamental studies of two-dimensional (2D) quantum-confinement phenomena and for property comparisons to 3D-confined quantum dots, 1D-confined quantum wells and anisotropically 3D-confined quantum rods. Potential applications of semiconductor nanowires in nanophotonics and lasing, nanoelectronics, solar-energy conversion and chemical detection are under active development. Exciting progress and advances in the semiconductor-nanowire field, enabled by the emergence of nanoparticle-catalyzed growth, are anticipated in the immediate future.

## Nanomaterials in Organic Photovoltaic Devices

The ability to create high-efficiency solar cells is a key strategy to meeting growing world energy needs. Nanotechnology is currently enabling the production of high-efficiency organic photovoltaics (OPVs) to help meet this challenge. Organic photovoltaics are nanostructured thin films composed of layers of semiconducting organic materials (polymers or oligomers) that absorb photons from the solar spectrum. These devices will revolutionize solar energy harvesting, because they can be manufactured via solution-based methods, such as ink-jet or screen printing, enabling rapid mass-production and driving down cost.

OPVs currently lag behind their "inorganic" counterparts because of low solar energy conversion efficiencies (approximately 1-3%). Several research groups are addressing conversion efficiency by employing a combination of nanomaterials and unique nanoscale architectures. These hybrid organic-inorganic photovoltaics consist of light-absorbing polymers in contact with semiconductor nanocrystals, fullerenes or nanostructured metals. The nanomaterials affect electro-optical properties of the conducting polymer, which include assisting in absorption of red and near-IR photons, a significant portion of the solar spectrum. Examples of OPVs designs employing nanomaterials include:

**Polymer-fullerene Heterojunctions**

Cells where a chemically modified fullerene ($C_{60}$) layer, acting as electron acceptors, is in close physical contact with a polymeric organic electron donor (MDMO-PPV catalog number 546461, or poly-(3-hexylthiophene), P3HT, catalog number 445703 and 510823). This contact improves efficiency by allowing charge transfer to take place at the sub 10-nanometer scale, on the order of the diffusion length of an exciton generated from organic semiconductors. The most recent cells exhibit conversion efficiencies of ~5%.

**Organic-nanocrytsal Solar Cells**

Blends of semiconducting polymers and semiconducting quantum dots or nanorods (CdSe or CdTe) are mixed in a manner similar to the polymer-fullerene blends. The polymers are modified to give rise to chemical bonding between the nanocrystal and polymer. The nanocrystals can be tailored to a wide variety of optical band gaps, which depend on the size of the nanocrystal (or the diameter of the nanorod).

**Dye-sensitized Cells**

These cells employ complex dye molecules attached to the surface of nanostructured oxides like titanium(IV) oxide ($TiO_2$) or niobium(V) oxide ($Nb_2O_5$). The dyes exhibit broad light-absorption profiles and rapid photoinduced charge transfer of electrons to the nanocrystals. These cells show solar conversion efficiencies of ~4%.

**Tandem Cells**

The tandem cell acts as a 2-1 cell, harvesting photons from the complete visible spectrum. These cells employ layers of $C_{60}$ as a strong blue light absorber and copper-phthalocyanine (CuPc) as a red-yellow absorber. Nanosized silver particles act as a charge conduit between the cells but do not absorb photons traveling through the cell because of their nanosized dimensions. These cells have achieved conversion efficiencies of ~6%.

Significant challenges exist to achieving OPV devices that can be mass-produced. Nanotechnology will assist in meeting the technical challenges of this rapidly evolving field.

## Coatings

Nanopowders and nanoparticle dispersions have seen increasing applications in coatings. Due to their small size, very even coating can be achieved by painting nanoparticle dispersions onto a surface and baking off the residual solvent.

### Optically Transparent Conductive Coatings

*Indium tin oxide* (ITO) and *antimony tin oxide* (ATO) are well known, optically transparent, electrically conductive materials. Nanoparticles of these materials can be painted on surfaces such as interactive touch screens to create a conductive, transparent screen without relying on expensive sputtering techniques. In addition, ITO and ATO can be used as an antistatic coating, utilizing their inherent conductivity to dissipate static charge.

### Optically Transparent Abrasion-resistant Coatings

Nanoscale aluminum oxide and titanium oxide are optically transparent and greatly increase the abrasion resistance of traditional coatings. Titanium oxide is of particular interest in many optical applications, since it is highly reflective for most ultraviolet radiation. Zinc oxide and rare-earth oxides are also UV-reflective, but optically transparent and are therefore effective in protecting surfaces from degradation brought about by exposure to UV radiation.

## Carbon Nanotube-Based Biosensors

### Bioconjugates

Researchers from the University of Illinois at Urbana-Champaign have developed near-infrared optical biosensors based on single-walled carbon nanotubes, which modulate their fluorescence emission in response to specific biomolecules. The viability of sensor techniques was demonstrated by creating a *single-walled carbon nanotube* (SWNT) *enzyme bio-conjugate* that detects glucose concentrations.

Carbon nanotubes fluoresce in a region of the near-infrared, where human tissue and biological fluids are particularly transparent to their emission. The sensor could be implanted into tissue, excited with a near-infrared light source, and provide real-time, continuous sensing of blood glucose level by fluorescence response. As shown in figure a schematic mechanism of the nanotube sensor. Hydrogen peroxide is produced when glucose reacts with the enzyme, which quickly transforms ferricyanide to modulate near-infrared fluorescence characteristics of the nanotube.

**Future Opportunities**

The important aspect of this technology is that the technique can be extended to many other chemical systems. New types of non-covalent functionalization are developed, creating opportunities for nanoparticle sensors that operate in strong absorbing media of relevance to medicine or biology.

## Layer-by-Layer Nanoassembly for Drug Delivery

Researchers from Louisiana Tech University were among the pioneers of one of the novel nanotechnology methods: *layer-by-layer* (LbL) nanoassembly by alternate adsorption of oppositely charged polyelectrolytes, nanoparticles and proteins. With this technique, we can assemble ultrathin multilayers with nanometer precision and pre-determined composition across the film, and make nanocapsules from LbL films. We are using such nanocapsules for targeted drug delivery, biocompatible nanocoating and pulp microfiber processing. For this nanoarchitecture, we use nanoblocks, such as nanoparticles. In the development of our research area, we successfully used nanoparticles, such as different diameter and surface-charged gold nanoparticles, silica, nanoclay-montmorillonite, alumina, titanium dioxide and other nanoparticles. Also, we investigated linear polyelectrolytes of different types and molecular weight (especially natural polyelectrolytes), which we used as electrostatic glue to assemble nanoparticle and protein arrays. As any architect, we nanoarchitects need a wide palette of nanoblocks with new properties and dimensions, ideally, monodispersed, stable in solution, charged nanoparticles of noble metals, metal oxides and ceramics with diameters of 5, 10, 20, 50 and 100 nm.

## Antibacterial Applications

Zinc oxide is an effective antibacterial and anti-odor agent. It has been used in deodorants, dental cleansers and diaper creams. The increased ease in dispersibility, optical transparence and smoothness make zinc oxide *nanopowder* an attractive antibacterial ingredient in many heath care products. Copper oxide nanopowder has also been proposed as an anti-microbial preservative for wood or food products.

## Nanomaterials in Fuel Cells

Over the last two decades, general interest and research in fuel cells has increased, because they have the potential to be more energy efficient than conventional power generation methods. During this time period, researchers have begun using nanomaterials in the catalyst layer of fuel cell electrodes for a variety of reasons, including: increasing the active surface area of the anode and cathode catalyst, increasing the catalytic rate of oxidation or reduction and minimizing the weight of platinum and other precious metals in the fuel cell.

The current generated at an electrode is proportional to the active surface of catalyst on the electrode surface, so higher power density fuel cells can be formed from nanomaterials, because nanomaterials have a higher surface area to volume ratio. Researchers have also shown that the electrocatalytic properties of the materials are sensitive to particle size, so increased catalytic activity can be observed for nanoparticles and nanomaterials.

However, the most important goal has been to decrease the weight of platinum and other precious metals in the catalyst layer of the fuel cell, so that the fuel cell can be cost-effective. This has been the main limitation to the widespread use of fuel cells. Researchers have employed carbon nanomaterials as supports for dispersions of platinum nanomaterials. This allows for a decrease in the weight of platinum needed to produce the same surface area of active platinum catalyst. The nanomaterials could be carbon foams containing nanopores, different types of nanotubes or even single-walled nanohorns. All of these materials can act as a support and a conductor for platinum nanomaterials, making strides toward cost-effective fuel cell catalysts.

## Nanomaterials in Other Energy Applications

### Environmentally Friendly Energy Sources

Devising schemes to meet the world's growing energy demands while simultaneously reducing green house emissions and other pollutants, has become one of the major challenges facing materials scientists. Nanomaterials promise to help solve many of the problems associated with new and emerging energy technologies.

#### *Fuel cells*

*Solid oxide fuel cells* (SOFCs) offer the advantage over other fuel cell designs in that they do not require expensive, precious metal catalysts and can operate effectively without extensive purification of fuel sources. The activity of doped rare-earth oxide electrodes such as yttrium stabilized zirconia (YSZ) is directly related to their surface areas. Nanoparticles exhibit the high surface required for developing SOFC technologies.

***Cleaner emissions***

Catalytic converters on vehicles around the world have significantly reduced the amount of automotive pollution over the last three decades. These devices require large amounts of expensive metals such as platinum, palladium and rhodium. Doped rare-earth metal oxides offer the promise of increased catalytic activity without the heavy reliance on precious metals. In addition, the increased efficiency of the next-generation catalytic converters will result in cleaner emission of existing internal combustion and diesel engines.

## Nanotube-Polymer Composites for Ultra Strong Materials

### Exceptional Strength

Exceptional mechanical properties of *single-walled carbon nanotubes* (SWNT) have prompted intensive studies of SWNT-polymer composites. However, the composites made with nanotubes are still holding a substantial reserve of improvement of mechanical properties. The problem is that pristine SWNTs have very poor solubility in polymers, which leads to phase segregation of composites. Severe structural inhomogeneities result in the premature failure of the hybrid SWNT-polymer materials.

The connectivity with the polymer matrix and uniform distribution within the matrix are essential structural requirements for strong SWNT composites. This problem is being solved by several approaches. First, by using coatings from surfactants and polymers, such as sodium dodecyl sulfate or poly(styrenesulfonate). This enables formation of better dispersions in traditional solvents including water. Polymeric dispersion agents are strongly preferred for the composite preparation because of (a) tighter bonding with the graphene surface, (b) miscibility with polymer matrixes of composites and (c) substantially smaller concentration necessary for the preparation of SWNT dispersions. Among polymers, different poly(vinyl alcohols) work best as a host matrix for SWNTs, providing the composites with high tensile strength and excellent Young's modulus.

### Looking Ahead

An additional resource of SWNTs for improving the mechanical properties of composites is nanotube orientation. Virtually complete alignment of nanotubes can be achieved in SWNT composite fibers. These composites display substantially better mechanical properties than any other SWNT-polymer hybrid. If one can find a simple and controllable method to produce not only fibers but also bulk materials and coatings with nanotubes oriented in a desirable direction, new technological vistas can be opened for various composites. Breakthroughs in this area can come both from the studies of fundamental properties of SWNTs and from development of methods for composite processing. In the past, magnetic field alignment with exceptionally powerful electromagnets and alignment in the flow was used for this purpose.

## Metal Matrix Nanocomposites for Structural Applications

### Improved Properties

*Metal matrix composites* (MMCs) such as continuous carbon or boron fiber reinforced aluminum and magnesium, and silicon carbide reinforced aluminum have been used for aerospace applications due to their lightweight and tailorable properties. There is much interest in producing metal matrix nanocomposites that incorporate nanoparticles and nanotubes for structural applications, as these materials exhibit even greater improvements in their physical, mechanical and tribological properties as compared to composites with micron-sized reinforcements. The incorporation of carbon nanotubes in particular, which have much higher strength, stiffness, and electrical conductivity as compared to metals, can significantly increase these properties of metal matrix composites. Nanocomposites are being explored for structural applications in the defense, aerospace and automotive sectors.

### Low Cost Solutions

Concurrent with the interest in producing novel nanocomposite materials is the need to develop low cost means to produce these materials. Most of the prior work in synthesizing nanocomposites involves the use of powder metallurgy techniques, which are not only high cost, but also result in the presence of porosity and contamination. Solidification processing methods, such as stir mixing, squeeze casting and pressure infiltration are advantageous over other processes in rapidly and inexpensively producing large and complex near-net shape components, however, this area remains relatively unexplored in the synthesis of nanocomposites. Stir mixing techniques, widely utilized to mix micron size particles in metallic melts, have recently been modified for dispersing small volume percentages of nanosize reinforcement particles in metallic matrices. Although there are some difficulties in mixing nanosize particles in metallic melts resulting from their tendency to agglomerate, a research team in Japan has published research on dispersing nanosize particles in aluminum alloys using a stir mixing technique. Researchers at the Polish Academy of Science have recently demonstrated the incorporation of greater than 80 volume percent nanoparticles in metals using high-pressure infiltration with pressures in the GPa range. Composites produced by this method possess the unique properties of nanosize metallic grains.

Recently, metal matrix nanocomposites were synthesized at the University of Wisconsin, Milwaukee using aluminum alloy A206 and nanoparticles of alumina ($Al_2O_3$). TEM samples of the cast Al-A206/$Al_2O_3$ clearly show nanoparticles present within the metal matrix. SAD patterns show the pattern of the matrix as well as the nanoparticles. EDX indicates that the grains are composed of aluminum, which contains nanosize alumina particles. The distribution of particles throughout the grains of the matrix with an absence of large concentrations at the grain boundaries suggests wetting of the alumina by the liquid metal. In this case the nanoparticles did not appear to act as nucleation sites for nanosized grains.

## Clusters and Nanoparticles

Particles with a size between 1 and 100 nm are normally regarded as nanomaterials. Figure 15.1 shows the size of nanoparticles in comparison with other small particles. In general, nanomaterials may have globular, plate-like, rod-like or more complex geometries. Near-spherical particles which are smaller than 10 nm are typically called clusters. The number of atoms in a cluster increases greatly with its

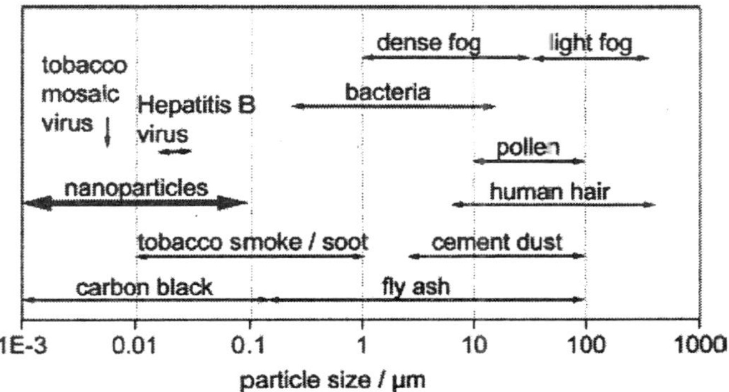


*Fig. 15.1. Typical size of small particles.*

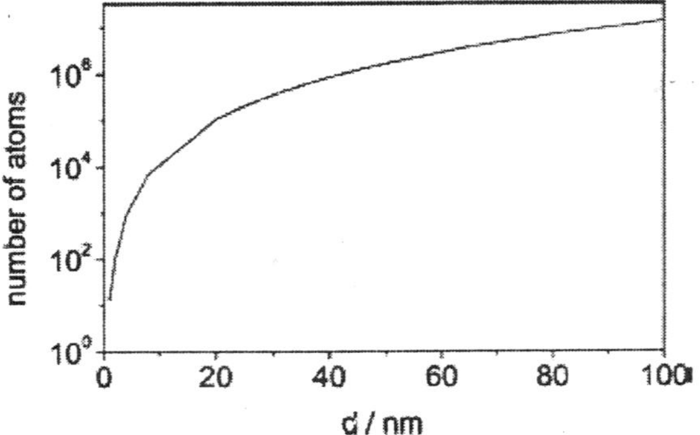


*Fig. 15.2. Number of sodium atoms in a spherical cluster of diameter d.*

diameter, demonstrated in Figure 15.2 for sodium clusters. At 1 nm diameter there are 13 atoms in a cluster and at 100 nm diameter the cluster can accommodate more than $10^7$ atoms. Clusters may have a symmetrical structure which is, however, often different in symmetry from that of the bulk. They may also have an irregular or amorphous shape. As the number of atoms in a cluster increases, there is a critical size above which a particular bond geometry that is characteristic of the extended (bulk) solid is energetically preferred so that the structure switches to that of the bulk.

It is below a dimension of 100 nm where properties such as melting point, colour (i.e. band gap and wavelength of optical transitions), ionisation potential, hardness, catalytic activity and selectivity, or magnetic properties such as coercivity, permeability and saturation magnetisation, which we are used to thinking of as constant, vary with size. We basically distinguish two types of variations as a function of size:

## Scalable Effects

Surface atoms are different from bulk atoms. As the particle size increases, the surface-to-volume ratio decreases proportionally to the inverse particle size. Thus, all properties which depend on the surfaceto-volume ratio change continuously and extrapolate slowly to bulk values.

## Quantum Effects

When the molecular electronic wave function is delocalized over the entire particle then a small, molecule-like cluster has discrete energy levels so that it may be regarded like an atom (sometimes called a super atom). The simplest model for it is that of a particle in a box. Adding more atoms to the cluster changes the size of the box continuously so that the energy levels close up to some extent. More importantly, adding more atoms means adding more valence electrons to the system. Thus, whenever a shell of sometimes multiple degenerate energy levels is .filled the next electron has to be accommodated in the next shell of higher energy. The situation is analogous to the evolution of properties with increasing atomic number in the periodic table. Filled shells represent a particularly stable configuration. Properties such as ionisation potential and electron affinity are well known to display a discontinuous behaviour as one moves along the periodic table. For clusters consisting of atoms with strongly overlapping atomic orbitals, i.e. for metals and semiconductors, the situation is analogous.

Quantum effects are more pronounced for small clusters and often superimposed on a smoothly varying background of a scalable effect. Clusters are interesting intermediates between single atoms and bulk matter and represent a natural laboratory to 'see both ends from the middle'.

## Feynman's Vision

On 29 December 1959, at the annual meeting of the American Physical Society, Richard Feynman addressed the audience with his visionary and by now historical and legendary lecture under the title – There is Plenty of Room at the Bottom: Invitation to Enter a New Field of Physics. With this talk on the problem of manipulating things on a small scale, Feynman opened the field of nanotechnology. Today, more than four decades later, the field is finally seen to really take off. It is amazing how closely some of the key developments follow Feynman's vision. Why cannot we write the entire 24 volumes of the Encyclopaedia Britannica on the head of a pin? he asked. We know how successful information technology has been in its work towards this goal and we should be aware of how much this has influenced our lives. Make the electron microscope hundred times better, Feynman said. The development of the atomic force microscope was one of the milestones on the way not only to observe but also to manipulate in atomic dimensions. Amazingly, Rohrer and Binnig achieved this goal with their cantilever-based instrument in a single step, rather than by a hierarchy of smaller and smaller robots – training an ant to train a mite – as Feynman suggested. In 1986, the two scientists were honoured for their achievement with the Nobel Prize in Physics. It is quite obvious today that the invention of the scanning tunnelling microscope finally triggered the boom in nanotechnology to which the direct observation of very small scale structures down to individual atoms is essential. Obviously, seeing things directly is more convincing for vision-based beings than just having measurements which are in agreement with a model.

Some of the inspiration came from biology. Feynman understood that information is stored on a molecular level in biology, that cells manufacture substances and operate on a small scale, that the human brain is a wonderful and efficient miniaturized computer. Consider the possibility that we too can make a thing very small which does what we want, an object that manoeuvres at that level, he suggested. While his talk was primarily technology oriented, he knew that physics, chemistry, biology and engineering are all relevant and must all be involved. He realized that making things smaller was not just a technological problem of scaling down. He saw that certain things changed principally. Magnetism, for example, is a cooperative phenomenon and involves domains which cannot be reduced down to an atomic size. Atoms differ from the bulk in their quantum nature. Most significantly, Feynman predicted that when we have some control of the arrangement of things on a small scale we will get an enormously greater range of possible properties that substances can have. It is exactly the arrangement of things on a small scale which is the foundation for all the excitement about nanomaterials and for the success of modern materials science.

# 16

# MICROCONVERSION

Machines, medicines, and materials a mere fraction the width of a human hair may one day store trillions of bits of information, detect the onset of cancer, and restore a paralyzed limb. But before the promise of nanotechnology is even modestly met, scientists must first learn how to make three-dimensional structures and tools at a scale much too small for even the deftest robots or nimblest human hands to manipulate.

The reduction in size of systems from computers to cell phones is a continuing evolution for electronic systems. The significance of this miniaturization goes well beyond just the smaller size and reduced weight. Batch fabrication, the parallel manufacturing of many integrated components, has been a key driver in the miniaturization of microelectronics because it reduces cost and increases reliability. Another significant trend is the integration of components and subsystems into fewer and fewer chips, enabling increased functionality in ever-smaller packages. These trends are extending to include micro-electromechanical systems (MEMS) and other technologies for sensors and actuators, thus allowing the possibility of miniaturizing entire systems and platforms. The combination of reduced size, weight, and cost per unit function has significant implications. Examples may include the rapid low-cost global deployment of sensors, launch-on-demand tactical satellites, distributed sensor networks, and affordable UAVs.

In the history of industrial engineering, technology characterized by length only occurred in microelectronics, but now we have nano-technology. How small is one nanometer? The typical width of a human hair is 50 micrometers. One nanometer is 50,000th of a hair width.

Nanotechnology is the construction and use of functional structures designed from atomic or molecular scale with at least one characteristic dimension measured in nanometers. Their size allows them to exhibit novel and significantly improved physical, chemical, and biological properties, phenomena, and processes because of their size. When characteristic structural features are intermediate between isolated atoms and bulk materials in the range of about one to 100 nanometers, the objects often display physical attributes substantially different from those displayed by either atoms or bulk materials.

Phenomena at the nanometer scale are likely to be a completely new world. Properties of matter at nanoscale may not be as predictable as those observed at larger scales. Important changes in behaviour are caused not only by continuous modification of characteristics with diminishing size, but also by the emergence of totally new phenomena such as quantum confinement, a typical example of which is that the colour of light emitting from semiconductor nanoparticles depends on their sizes. Designed and controlled fabrication and integration of nanomaterials and nanodevices is likely to be revolutionary for science and technology.

Nanotechnology can provide unprecedented understanding about materials and devices and is likely to impact many fields. By using structure at nanoscale as a tunable physical variable, we can greatly expand the range of performance of existing chemicals and materials. Alignment of linear molecules in an ordered array on a substrate surface (self-assembled monolayers) can function as a new generation of chemical and biological sensors. Switching devices and functional units at nanoscale can improve computer storage and operation capacity by a factor of a million. Entirely new biological sensors facilitate early diagnostics and disease prevention of cancers. Nanostructured ceramics and metals have greatly improved mechanical properties, both in ductility and strength.

From the fundamental units of materials, all natural materials and systems establish their foundation at nanoscale; controlling matter at atomic or molecular levels means tailoring the fundamental properties, phenomena, and processes exactly at the scale where the basic properties are initiated. Nanotechnology could impact the production of virtually every human-made object—from automobiles and electronics to advanced diagnostics, surgery, advanced medicines, and tissue and bone replacements. To build electronic devices using atom-by-atom engineering, for example, we have to understand the interaction among atoms and molecules, how to manipulate them, how to keep them stable, how to communicate signals among them, and how to face them with the real world. This goal requires new knowledge, new tools, and new approaches.

## Much More than Miniaturization

To many people, nanotechnology may be understood as a process of ultraminiaturization. Philosophically, changes in quantity result in changes in quality. Shrinkage in device size may lead to a change in operation principle due to quantum effect, which is the physics that governs the motion and interaction of electrons in atoms. In fact, the trend in product miniaturization will require new process measurement and control systems that can span across millimeter-, micrometer-, and nanometer-sized scales while accounting for the associated physics that govern the device and environment interaction at each specific size scale.

To consider the interactions among atoms in the nanometer scale, we need to introduce quantum mechanics and each atom has to be treated as a unit. To face the atoms with the real world in the millionmeter scale, we need to consider the collective properties of millions and millions of atoms, so that the matter is considered to be a continuous medium, and we use classical mechanics. The bridging of the two length scales requires new standardized architecture definitions that support multiple physics-based models and new computational representations that allow seamless transition and traversing through these various models.

### Manufacturing Nanomaterials

Nanomanufacturing technologies that will support tailor-made products having functionally critical nanometer-scale dimensions are produced using massively parallel systems or self-assembly. The current research mainly focuses on nanoscience for discovering new materials, novel phenomena, new characterization tools, and fabricating nanodevices. The future impact of nanotechnology to human civilization is manufacturing. The small feature size in nanotechnology that limits application of well-established optical lithography and manipulation techniques causes industrial nanomanufacturing to remain a serious challenge to our technological advances.

Synthesis of nanomaterials is one of the most active fields in nanotechnology. There are numerous methods for synthesizing nanomaterials of various characteristics. An essential challenge in synthesis is controlling the structures at a high yield for industrial applications. Techniques are needed for atomic and molecular control of material building blocks, which can be assembled, used, and tailored for fabricating devices of multi-functionality in many applications.

The oxide nanobelt discovered in my laboratory is an example. Ultra-long nanobelts have been successfully synthesized for a wide range of oxides by simply evaporating the desired commercial metal oxide powders at high temperatures. These materials are semiconductors with important applications in sensors and transducers. The as-synthesized oxide nanobelts are pure and structurally uniform; they have a rectangular-like cross-section. The semiconducting oxide nanobelts could be doped with different elements and be used for fabricating nanometer-sized sensors based on the characteristics of individual nanobelts, which could be potentially useful for in-situ, real-time, and remote detection of molecules, cancel cells, or proteins based on electronic signal. The nanobelts could also be used for fabrication of nanoscale electronic and optoelectronic devices because they are semiconductors.

**Characterizing the Performance and Properties of Nanostructures**

Property characterization of nanomaterials is challenging because of the difficulties in manipulating structures of such small size. New tools and approaches must be developed to meet new challenges. Due to the high size and structure selectivity of nanomaterials, their physical properties could be quite diverse, depending on their atomic-scale structure, size, and chemistry. A typical example is the carbon nanotube, which is made of concentrical cylindrical graphite sheets with a diameter range from one to 400 nanometers and length of a few micrometers. Characterizing the mechanical properties of individual nanotubes, for example, is a challenge to many testing and measuring techniques because of the following constraints. First, the size (diameter and length) is rather small, prohibiting the application of well-established testing techniques. Tensile and creep testing require that the size of the sample be sufficiently large to be clamped rigidly by the sample holder without sliding. This is impossible for one-dimensional nanomaterials using conventional means. Second, the small size of the nanostructure makes their manipulation rather difficult, and specialized techniques are needed for picking up and installing individual nanostructures. Therefore, new methods and methodologies must be developed to quantify the properties of individual nanostructures.

In-situ transmission electron microscopy technique, or TEM, has been developed for measuring the modulus of individual carbon nanotubes. We have to see the object while its properties are being measured, thus, a microscope is required. To carry out the property measurement of a nanotube, a specimen holder for a TEM was built for applying a voltage across a nanotube and its counter electrode. To measure the bending modulus of a carbon nanotube, an oscillating voltage is applied on the nanotube that can tune the frequency of the applied voltage. By changing the frequency of the applied voltage onto the nanotube, mechanical resonance can be induced in carbon nanotubes at specific frequencies from which the bending modulus can be derived. This type of technique works well for small objects.

**Large-scale Manipulation and Self-assembly**

Manipulation of nanostructures relies on scanning probe microscopy. Using a fine tip, atoms, nanoparticles, or nanowires can be manipulated for a variety of applications. This type of approach is outstanding for scientific research. For manufacturing, an array of scanning tips, if synchronized, may be used for achieving atom-by-atom engineering. But the building rate is rather slow. If a device has a feature size of five nanometers and a scanning tip can move atoms 10 atoms per second, it will take about six months to build $10^{12}$ devices on an eight-inch wafer.

The ultimate solution is self-assembly. Like many biological systems, self-assembly is the most fundamental process for forming a functional and living structure. The genetic codes and sequence built in a biosystem guide and control the self-assembling process. Self-assembly is the organization and pattern formed naturally by the fundamental building blocks such as molecules and cells. Designed and controlled self-assembly is a possible solution for future manufacturing needs.

Size- and shape-selected nanocrystals behave like molecular matter that can be used as fundamental building blocks for constructing nanocrystal-assembled superlattices. Self-assembled arrays involve self-organization into monolayers, thin films, and superlattices of size-selected nanocrystals encapsulated in a protective compact organic coating. Nanocrystals are the hard cores that preserve the ordering at the atomic scale; the organic molecules adsorbed on their surfaces serve as the interparticle molecular bonds and as protection for the particles in order to avoid direct core contact with a consequence of coalescing. The interparticle interaction can be changed via control over the length of the molecular chains, resulting in tunable electronic, optical, and transport properties.

### Large-scale Parallel Device Fabrication and System Integration

System integration involves at least an integration of numerous functional materials and components for achieving a complex, preprogrammed action. This involves patterned materials growth on a designed substrate; large-scale, parallel integration of nanowires, nanoparticles, and functional groups; interconnection among the components; and defect-tolerated path design following neuron networks. A wide range of novel approaches has been developed for fabrication of single nanodevices. Nanomanufacturing requires a simultaneous, parallel fabrication of a large amount of nanodevices under precisely controlled conditions and repeatability. This remains a major challenge to the development of nanotechnology, especially for nanoelectronics. A possible solution is to integrate patterns produced by lithographic technique with self-assembly process. Self-assembly of single-walled carbon nanotubes is one example.

By functionalizing the substrate produced by lithographic technique so that individual carbon nanotubes selectively recognize the locations for self-assemble on the substrate following specific patterns, mass producing carbon nanotube-based circuit structures is possible. To achieve the process, the substrate was coated with patterns of organic molecules using techniques such as dip-pen nanolithography and microcontact stamping. Two surface regions have been produced: one patterned with polar chemical groups, and the second coated with nonpolar groups. A suspension of single-walled carbon nanotubes solution was added, and the nanotubes were attracted to the polar regions and self-assembled to form predesigned structures. Millions of individual nanotubes have been patterned on stamp-generated microscale patterns, covering areas of about one square centimeter on gold.

### Integrating Nanometer-to-millimeter Manufacturing Technologies

Over the next decade, major industrial and scientific trends that emerged during the 1990s will influence not only how manufacturing will be done, but also what is manufactured. The size of many manufactured goods continues to decrease, resulting in ultraminiature electronic devices and new hybrid technologies. For example, micro-electromechanical system, or MEMS, devices integrate physical, chemical, and even biological processes in micro- and millimeter-scale technology packages. MEMS devices are used in many sectors: information technology, medicine and health, aerospace, automotive, environment, and energy, to name a few. The future relies on the integration of nanotechnology with existing technology.

The challenge remains in integrating nanotechnology with microelectronics-based technology. The nanometer-scale components have to be connected with micrometer- and millimeter-scale components to communicate with the real world. This requires an integration of not only the technologies covering nanometer- to-millimeter multilength scales, but also the physics and chemistry covering the entire length scale. We are facing the merging of quantum mechanics and classical mechanics. Any ultra small components have to be connected with the real world. The goal should be on how to use nanotechnology to make microtechnology more efficient, multifunctional, and intelligent as well as faster, smaller, and achieving the impossible. Nanotechnology comes to life if we can achieve the integration of nanoscale building blocks with lithographically produced structures through self-assembly and genetically engineered growth.

**Building Nanomanufacturing Standards**

Nanomanufacturing needs the measurements and standards required to achieve effective and validated nanoscale product and process performance. This challenge is mainly in the following three directions:

***Atomic-scale manufacturing***

Develop and assemble the technologies required to fabricate standards that are atomically precise. This will include work directed at solving artifact integrity, precision placement, dimensional metrology, and manufacturing issues.

***Molecular-scale manipulation and assembly***

Identify and address the fundamental measurement, control, and standards issues related to manipulation and assembly of microscale or nanoscale devices using optical, physical, or chemical methods. This entails building the manipulation technology and using it to understand and address the measurement issues that arise when assembling devices at the microscale or nanoscale level.

***Micro-to-millimeter-scale manufacturing technologies***

Develop the technologies required to position, manipulate, assemble, and manufacture across nanometer-to-millimeter multilength scales.

**Not Just an Engineering Process**

Traditionally, manufacturing is attributed to an engineering field. For nanomanufacturing, we must go beyond engineering. Once we approach the atomic-scale precision and control, fundamental physics and chemistry have to be applied. The nanoscale manufacturing is multidisciplinary—involving but not limited to mechanics, electrical engineering, physics, chemistry, biology, and biomedical engineering. The future view of nanomanufacturing is the integration of engineering, science and biology. This complex task requires not only innovative research and development themes, but also a new education system for training future scientists and engineers.

**New Engineered Materials**

Advances in micro- and nanofabrication technologies are enabling the engineering of materials down to the atomic level. While design and fabrication capabilities are still primitive from an applications perspective, there is great potential for improving the properties and functionality of materials. Examples of recent advances in materials range from carbon nanotubes with great strength and novel electronic properties, to quantum dot communication lasers, to giant magnetoresistive materials for high-density magnetic memories. Theory and simulation will play an increasingly important role in guiding the development of new nanostructured materials and of systems based on such materials. By combining materials at the micro- and nanoscales to form smart composite structures, additional increases in functionality can be achieved. New materials are an underlying enabling capability. They will be used to expand the performance envelope of electronics, sensors, communications systems, avionics, air and space frames, and propulsion systems. Theory and modeling of materials are advancing significantly, as is our understanding of the relationships between material composition, properties, and structure. Over time, these advances may reduce the long lead time for developing new materials and may also help in the design of new, more functional materials.

**Increased Autonomy and Functionality**

The advances in information density, miniaturization, and materials functionality will enable a degree of autonomous operation for systems that cannot be fully envisioned today. Enhanced functionality and increased autonomy based on micro- and nanotechnologies have many systems benefits: (i) lower risk for humans, (ii) higher performance, (iii) lower cost platforms, and (iv) reduced communication requirements with a correspondingly lower probability of detection

Initially, this autonomy will be seen simply as an evolutionary extension of the capabilities of current systems, providing increased accuracy and range or other performance advantages. Over the longer term, however, the dramatic increases in local information awareness and computational power will enable independent decision making and will have a dramatic impact. Systems may also be able to power, self-repair, and reconfigure themselves to extend the scope of their missions. The lowered cost and increased functionality will lead to swarms of intelligent agents with emergent behaviour that differs from that of any single entity. Integrating these advances will be challenging and will raise important global political and societal issues as well.

## Device Miniaturization

Miniaturization of devices is another area that will rely on biology and computing as it offers exciting possibilities in medicine, energy efficiency, national security, and environmental protection. Al Romig of Sandia National Laboratories reported that miniaturization will proceed both by making today's components smaller and by taking micro-scale materials, including biological materials, and building up from there. Thus, he said, not only will the physical sciences help biomedical research, but also biological research will lead to advances in the physical sciences. He is confident that over the next 10 to 20 years, a growing number of materials and devices on the market will be manufactured using biological methods.

In the field of nanotechnology, for example, nanocrystals have been injected into living cells. These crystals do not damage cells. Rather, they emit light enabling scientists to explore further the properties and behaviour of cells. Nanotechnology will also enable the creation of more efficient materials, which should reduce energy consumption with major benefits for the environment.

Nanoelectronics and micro-electromechanical systems (MEMS) will enable electronic systems to carry out essential battlefield functions that humans do now at their peril. For example, micro-air vehicles are being developed to jam enemy radar systems, and biosensors will detect the presence of dangerous substances in the environment. As battlefield requirements increase, the military will look to biological systems for solutions, but developing those technologies will rely on advances in computing and electronics. In turn, this will require the confluence of a variety of research disciplines, including the biosciences, materials science, computers and information technologies, and electronics.

## Miniaturization and Large Scale Integration

After the invention of the integrated circuit, it took very little time to realize the tremendous benefits of miniaturizing and integrating larger numbers of transistors into the same integrated circuit. More transistors (switches) were required in order to implement more complicated functions. Miniaturization was the key to integrating together large numbers of transistors while increasing hardware speed and keeping power consumption and space requirements manageable.

*Large scale integration* ("LSI") came to refer to the creation of integrated circuits that had previously been made from multiple discrete components. These devices typically contained hundreds of transistors. Early computers were made from many of these smaller ICs connected together on circuit boards.

As time progressed after the invention of LSI integrated circuits, the technology improved and chips became smaller, faster and cheaper. Building on the success of earlier integration efforts, engineers learned to pack more and more logic into a single circuit. This effort became known as *very large scale integration* (VLSI). VLSI circuits can contain millions of transistors.

Originally, the functions performed by a processor were implemented using several different logic chips. Intel was the first company to incorporate all of these logic components into a single chip. This was the first microprocessor, the 4004, introduced by Intel in 1971. All of today's processors are (highly advanced!) descendants of this original 4-bit CPU.

**Hard-wired vs. Programmable Logic**

There is one primary distinguishing feature of a processor that makes it more than just a highly-integrated regular hardware circuit. Ordinary electronic hardware implements a predefined set of functions; the logic is "hard wired" for a single purpose, and the logic circuits are customized to that task only. A different use would require changes to the circuitry.

Processors are *programmable*, that is, they can perform different functions based on instructions read from a program. Instead of being customized to one single task, a processor is designed in a more general way to perform a broad range of functions. It does this by defining a set of subtasks, or *instructions* that controls how it works, and then letting users write programs from these instructions. This design introduces tremendous flexibility over the hard-wired design (although it comes at the small penalty of being slightly less efficient than a customized machine.)

All modern computers use this "hardware-software" model, where hardware performs functions under the control of a program, which is of course called *software*. In fact, modern computers are made of many layers of software and hardware, each controlled by programs written at the next higher layer.

**Logic Design**

The actual design of the microprocessor is done first at a logical level. The processor is defined in terms of its desired operation, what instructions it is meant to operate on, and how data is intended to flow between its various logical parts. This high-level design starts at a very conceptual level, and proceeds to more detailed levels where the various architectural features of the chip are defined in great detail, such as how instructions will be decoded, how the control unit(s) will function, etc. This is a very abbreviated description of the process, which involves hundreds or even thousands of engineers and takes months or years of time depending on how advanced the device is and also on how much is borrowed from previous designs.

Once the processor has been fully defined in this manner it is laid out physically. The actual transistors necessary to implement the design are determined and a physical map is created showing how the chip must be fabricated. In real life there isn't just one design that is then mapped out in this way once; it is an iterative process with changes being made all the time in the design, due to bugs, enhancements, marketing issues, or manufacturing concerns.

**Photolithography**

Once the design of the processor has been completed (or at least a current version of it) and the low-level transistor maps have been defined, it is necessary to actually make the chips. The chips are manufactured using a process called *photolithography*. In simple terms, transistors and data pathways (internal wires) are created in semiconductors by applying different layers of various materials at exact places, one after the other. Where two layers intersect in the appropriate way, a transistor is formed. A modern processor can be made of a dozen or more of these layers.

A special (very expensive) computer system takes the transistor design of the processor and separates it into these different layers, each of which is called a *mask*. This name comes from the way the chip is actually made. A wafer is chemically treated with a photosensitive substance, and then light is shined through the first layer mask onto the wafer. Where the mask allows the light to shine onto the wafer, it is chemically altered, creating the first layer of the chip. Then another chemical is used to wash off the first one, and the process is repeated with the next mask. In this way, the features of the chip are slowly built onto the silicon, layer by layer. In order to achieve the miniaturization that gives the processor its great power in a small size, the features (transistors and internal wires) of the processor are shrunk to a microscopic size—less than a micron in width, or one millionth of a meter. Each mask must be created with drawn features that are this size, and the device that does the photolithography must be able to focus light at this level of accuracy as well. Obviously, the equipment to perform this work is amazingly precise, and incredibly expensive.

As you recall, the wafers used in this process are cut from a large crystal, typically eight inches in diameter. Each chip is very small, usually a third of a square inch or less, so each wafer contains many hundreds of chips. The wafer is essentially "chopped up" into little rectangles, each one of which has the potential to become a chip. Some of the surface area of the wafer is also used for test circuits. When the wafer is complete, another special machine breaks it into its individual chips.

**Circuit Size**

The circuit size or feature size refers to the level of miniaturization of the processor. To make more powerful processors, more transistors are needed. In order to pack more transistors into the same space, they must be continually made smaller and smaller. As processors get faster and denser, power consumption and heat generation become a big concern as well. Circuit size is a major limiting factor in processor speed, largely due to heat generation issues. It also has a major impact on die size.

Technology advancements continue to allow circuit sizes to shrink and shrink. It was once considered "impossible" to shrink the circuit size below 1 micron; most processors today actually now use a 0.35 micron process, with 0.25 the current state of the art in the latest chips and 0.18 micron processors looming in the near future. It is now thought that current technology can eventually be shrunk to as low as 0.08 microns. This will make possible a whole new generation of very fast, low-power devices.

**Die Size**

The die size of the processor refers to its physical surface area size on the wafer. It is typically measured in square millimeters (mm^2). In essence a "die" is really a chip, but it is only referred to in this way when discussing physical chip parameters and manufacturing issues. The importance of die size is rather obvious: the smaller the chip, the more of them that can be made from a single wafer. A larger die means fewer chips from the same wafer, and thus higher cost overall. A larger die also leads to increased power consumption. The three most important contributing factors to die size are the circuit size in microns, the process technology used, and of course, the design of the processor itself (newer processors are in general larger because they do a lot more). Reducing circuit size in particular is key to reducing the size of the chip. For example, the first generation Pentium used a 0.8 micron circuit size, and required 296 square milimeters per chip. The second generation chip had the circuit size reduced to 0.6 microns, and the die size dropped by a full 50% to 148 square milimeters.

Improving the ability to control matter has long been a major aim of technology. Progress in science suggests the feasibility of achieving thorough control of the molecular structure of matter via controlled molecular assembly. The consequences of such an ability will be enormous, in areas as diverse as medicine, computation, manufacturing, and the environment. We can gain some understanding of these future possibilities by observing examples from nature.

Molecular assembly in nature operates by means of molecular machines: machines which use individual atoms and molecules as working parts. Enzymes are molecular machines that make, break, and rearrange chemical bonds at a rate of up to a million per second. Muscle fibers work like molecular-scale linear motors. DNA serves as a digital storage medium, directing ribosomes in manufacturing proteins. Nanotechnology will likewise use programmable molecular machines to build complex structures atom by atom.

The term nanotechnology has sometimes been used to refer to any technology giving some control at a substantially submicron scale. Here the term is used more narrowly to describe a technology giving more general control of the structure of matter on a nanometer scale: a broad ability to control the arrangement of atoms.

## Top-Down Approach

An approach to atomic-scale control was outlined in 1959 by Richard Feynman. He suggested that large machines would be used to build smaller machines, which would build yet smaller machines, working toward molecular dimensions. This sequence has been termed a "top-down" strategy.

Modern microtechnology uses a different top-down approach—that of photolithography, thin films, and selective etching—to make microelectronic systems. These techniques are now being used to make mechanical structures such as gears and motors. Currently, such devices are tens of microns in diameter, many orders of magnitude larger than nanometer-scale.

Chemists, in contrast, have started at the atomic level, building molecules precisely. Progress in molecular biology and synthetic chemistry has been steady and impressive: the operation of many naturally-occurring molecular machines has been elucidated, and some have been synthesized. This work, led by K. Eric Drexler, is to propose a "bottom-up" approach to artificial molecular machines, ultimately including devices such as molecular assemblers, assembler-based replicators, and mechanical nanocomputers. The "bottom-up" style builds toward larger and more complex systems by starting at the molecular level and maintaining precise control of molecular structure. One approach would involve designing new protein-based devices which self-assemble to form molecular machines like those found in nature. Natural molecular devices show that such machines could include parts acting as struts, cables, fasteners, motors, bearings, and numerical control systems.

**Current Work**

Non-molecular machines may also be of use in bottom-up strategies. The scanning tunneling microscope (STM), announced by Binning and Rohrer in 1982 and for which the Nobel Prize in Physics was awarded in 1986, and the atomic force microscope (ATM) announced by Binning and Quate in 1986 are both capable of positioning a sharp tip with atomic accuracy. These tips might be equipped with molecular tools capable of carrying out specific reactions. However, performing only one molecular reaction at a time will be impractical for making large amounts of a product. It appears that until assemblers capable of replication can be built, the parallelism of chemical synthesis and self assembly will be needed; groups of molecules can self assemble quickly due to their thermal motions, which enable them to "explore" their environments and find (and bind to) complementary molecules.

Given their key role in natural molecular machines, proteins are obvious candidates for early work in self assembling artificial molecular systems. Recent work by DeGrado shows the feasibility of designing protein chains that fold predictably into solid molecular objects. Progress is also being made in artificial enzymes and other relatively small molecules that perform functions like those of natural proteins; the 1987 Nobel Prize in Chemistry went to Cram and Lehn for such work. Several bottom-up strategies using self assembly seem feasible.

## Intermediate Goals

Bottom-up strategies are being pursued in academic and industrial labs because molecular engineering has short-term rewards. Improved enzymes, sensors, and detectors are obvious applications. More ambitious is work in molecular electronics, aimed at constructing circuitry in which individual molecular parts serve as wires and transistors.

But before reaching this stage, the bottom-up approach should yield gradually more complex molecular machinery, perhaps starting with folding polymer systems able to help synthesize copies of themselves. These could lead to crude "proto-assemblers" (which might or might not incorporate an AFM or STM positioning mechanism) able to manipulate a limited set of molecular tools. With these devices, building an actual assembler with broad capabilities should be within reach.

## Assemblers and Nanotechnology

General-purpose assemblers will use chemical reactions combined with precise location control to produce materials not accessible to conventional synthetic chemistry. Products made by assembler could include many mechanical parts on a molecular scale: drive shafts, cams, roller bearings, levers, even electrostatic motors. Describing potential products of nanotechnology involves the discipline of "exploratory engineering": the study of what can be built with tools that are foreseeable, but not yet available. Exploratory engineering aims not to predict what will be built, but to make a sound case that certain kinds of systems can be built. This favours simple, easily-analyzed systems.

Molecular electronic computers built with nanotechnology seem feasible, but are difficult to design. In contrast, molecular mechanical computers—molecular Babbage machines—can be designed and analyzed more easily. Design calculations indicate that a mainframe computer built with this technology could have a volume as small as a cubic micron. Systems incorporating nanocomputers, assemblers, and suitable programs and auxiliary devices should be able to build copies of themselves as cheaply as bacteria do. Thus, assemblers and their products can eventually become inexpensive.

## Applications

With replication, assemblers will be able to work in parallel to produce objects of macroscopic size. Special purpose devices could be designed to tackle such enormous tasks as removing excess carbon dioxide from the atmosphere or to produce goods for human consumption. Perhaps most intriguing are the possibilities for radically improved medical care, combining the advantages of molecular action of drugs with those of surgical control.

Almost any technology can be abused, and nanotechnology will be no exception. A technology for inexpensive mass production could be used to produce weapons, and worse scenarios (e.g. programmable "germs" for germ warfare) have been discussed. Nanotechnology will let us control the structure of matter—but who will control nanotechnology? The chief danger isn't a great accident, but a great abuse of power. In a competitive world, nanotechnology will surely be developed. If democratic institutions are to guide its use, it must be developed by groups within their political reach. To keep it from being developed under a shroud of military secrecy, it seems wise to emphasize its value in medicine, the economy, and in restoring the environment. Nanotechnology must be developed openly to serve the general welfare.

## Miniaturization in Metal Wires

Dabs of metal too small to be seen could start the next revolution in electronics.

The smaller the wires and switches on those silicon chips that run everything from talking teddy bears to supercomputers, the more efficient and less expensive these things become. Since the 1960s the number of switches and other elements on a fingernail-size chip has doubled every 18 months from dozens to tens of millions. But there's got to be a limit.

The new field of nanotechnology can now produce wires with diameters a scant few hundreds of millionths of an inch in diameter, a hundred or so atoms across. You can't even see wires that small; you need a sophisticated electron microscope.

What keeps such state-of-the-art wires out of laptops and cell phones is contacts. Contact points on the switches and other circuit parts are chunks of metal hundreds of times bigger than nanowires that could connect them together. In a basement laboratory at Harvard University, Charles Lieber and his students have solved this problem. They have taken the first step toward making chips with billions instead of millions of components, nanoelectronics instead of microelectronics.

"Nano" refers to nanometers, one billionth of a meter, four hundred millionths of an inch, hundreds of thousands of times thinner than a hair, about 10 small atoms in width. Using chemical processes developed at Harvard's Department of Chemistry and Chemical Biology and Division of Engineering and Applied Sciences, Lieber and his team have made wires as thin as 3 nanometers, tens of atoms thin. What is more, the nanotechnologists have built switches right into the wires, solving the tedious problem of connecting the switches, amplifiers, and other devices that, in today's integrated circuits, are so much bigger than the nanowires.

Such integrated circuits could have applications well beyond faster, smaller computers and cell phones with features only fantasized about today. For example, nanocircuits might make possible sensors that can detect a single virus in your blood.

It could turn manufacturing of high-end technology upside down. It could affect all electronic circuits in the world.

Semiconductors can, under different conditions, conduct or block electric current, making them ideal switches. But does the behaviour of electric currents change when metal junctions shrink to atomic levels? That's what we need to understand in order to develop entirely new technologies, technologies we cannot even predict today.

Present methods of circuit making involve the etching of semiconductors and metals with energetic beams of X-rays or electrons. As switching devices like transistors get smaller and smaller, their behaviour starts to worsen. Edges get rougher, less well defined, and electric currents are slowed as they try to move through different circuit components. Lieber's team uses chemical assembly, evaporating metal coatings on top of silicon nanowires, and then knocking off excess metal with chemical etching. Lieber describes wires and transistors made this way as "essentially perfect. They show enhanced behaviour at these vanishingly small scales."

At the outset, it was not obvious that such a thing would work. The different materials would combine in a way that produced wires containing too many defects to be useful. That would be true with normal-size elements, but at nanoscales, things came together smoothly. There are no defects between the different materials that would disrupt electric currents.

Contrary to what you might think, atomically thin nanowires can carry currents a hundred times greater than those thin copper wires we are all so familiar with. Best of all, Lieber doesn't think nanowires and transistors made his way will be any more expensive than microdevices used today.

## Bioinspired Miniaturization

From previous results attained on the basis of thermodynamics, statistical physics and nanobiology it is shown that Artificial Intelligence (Al) concepts and their hard-wired implementations cannot satisfy the physical conditions for mimicking biological intelligence and its capabilities for learning, which actually is a self-organization process. Moreover, biosystems feature a structure-function solidarity, i.e. an indistinguishability of hardware and software which, as shown, asks for integration of the microphysical and the macroscopic world, to be carried out going through the nanoscale (quantum/classical) physics. A highly nonlinear, far-from-equilibrium, dissipative hierarchical dynamics is shown to implement such conditions for a bioinspired miniaturization aiming at intelligence metaphors closer to biological intelligence than those attainable through Al. It is shown that the mathematical techniques of the emerging science of Quantum Holography (QH) can be applied to Nanotechnology through proper modeling of the nilpotent Heisenberg's Lie group G of symmetry, which is shown to analyze and synthesize the convolution structure of the wavelets originating from a mother wave in the case of phase coherence under the action of Fourier transform. Such symmetry avoids quantum decoherence at the microlevel, so that a system behaviour stretching microlevel quantum coherence information up to nano/macrolevel dissipation information can be attained. This nano-to-micro integration biomimetics, applied to nanophotonics, microoptics, MOEMS and BioMEMS is shown to realize the closed chain of sensing-information processing-actuating that embodies the biological structure-function solidarity and features no algorithmic instructions, but just learning; no net hardware-software distinction, but just a physical morphological solidarity. QH is also developed into a Generalized QH that applies to any kind of wave phenomenon, so opening the way to systems other than photonic.

# 17

# DEFENSIVE NANO

The dramatic improvements in information technology over the last 50 years have led to a revolutionary change in the conduct of warfare. Information superiority is identified by the Air Force as a Core Competency, critical to modern warfare. The Critical Future Capability statement demands that "continuous, tailored information be provided within minutes of tasking with sufficient accuracy to engage any target in any battle space worldwide." Information superiority is a critical component of other Core Competencies, including Aerospace Superiority, Global Attack, and Precision Engagement. Each of these is critically dependent on accurate and timely information.

Several pieces must come together to satisfy the Air Force's requirements. Sensors, discussed later, provide the raw data. Electronic signal processing is applied to the sensor outputs to interface with the larger-scale information processing and communication systems. Communication at many levels is necessary to gather the information. Information processing, including fusion of data from multiple sensors, distills the sensor data into the information necessary for decision making. At each step there are requirements for data storage and display as well as for computation. Ultimately this must be a robust, redundant system tolerant of the failure of individual segments and self-reconfigurable to adjust to changing conditions and demands.

Research and development (R&D) investments by the Air Force in these areas must be considered in context with other worldwide efforts, especially those of industry. Because the DoD is today a relatively small customer for information technology, industrial R&D programs are considerably larger than those affordable by the Air

Force. Clearly, the commercial sector provides immense incentive for going after scientific and technological advances in this field. However, some technologies that are unique to the military or that are not yet commercially viable require military investment:

... For example, many sensor applications are unique to government requirements and hence are funded solely by the government. Similarly, there are additional technologies that are essential for government missions but which may have or develop commercial application as well; however, the cost of their development is usually so high that industry cannot make a business case for maturing them commercially. Examples include the Global Positioning System, or development of new propulsion concepts.

Assessing the appropriate R&D investment by the Air Force in light of the ongoing revolution in information technology in the commercial sector is challenging. In this section we examines some of the specific sectors of information technology (IT) to draw distinctions for the Air Force.

This section starts with computing devices. Transistors, switches, and integrated circuits are covered; a separate section on space electronics is included because of the unique environmental requirements of space and its importance to the Air Force. Also covered are storage and display technologies. Computing architectures explores alternative paradigms for computing. Under communications, a number of areas are explored: optical materials and devices, radio frequency (RF) materials and devices; and RF and optical MEMS. Finally, information and signal processing and data fusion requirements are discussed.

## Devices

### Scaled Metal Oxide Semiconductors

Advances in information technology are a result of the ever-shrinking transistor, applied to almost every aspect of gathering and treating information. Continuing increases in information technology capabilities are dependent on continuing advances in the fabrication of ever more powerful computational hardware. Moore's law, the exponential increase in integrated circuit functionality, continues today. Although there are potential limitations on the horizon, the semiconductor industry roadmap, ITRS, which is based on the scaling from larger devices that has served us so well for the last 40 years, foresees a continuation of the current rate of Moore's law to the present roadmap horizon of 2016, corresponding to a 32-fold improvement in device density. Barriers to reaching and surpassing this density include the following, among others: lithography, gate oxide current leakage, interconnect requirements, and thermal issues. These challenges have spurred many research efforts, both to address the issues within the context of traditional scaled-silicon systems and to find alternatives that circumvent the approaching barriers. Of great interest has been the field of nanotechnology, where multiple materials are innovatively positioned with nanometer precision.

The ITRS, as discussed is industry's best analysis of all factors (fabrication, interconnects, thermal management, cross talk, cost, packaging, etc.) that must be considered in order to continue the miniaturization progress. The ITRS identifies a number of "brick walls"—major technology issues for which there is no known solution—as well other important technological challenges for which promising approaches have been identified. Given the current state of knowledge of CMOS and of the alternatives as they are now understood, the best guess for the next 10-15 years is that silicon CMOS technology will continue to provide the fastest switching time at the lowest cost in the smallest gate with the most cost-effective system integration.

The continuing hegemony of CMOS devices and circuits is based on substantial improvements and the introduction of highly innovative ideas. At the 2001 International Electron Devices Meeting, Intel announced a transistor operating at 3.3 terahertz. At this same meeting. Advanced Micro Devices (AMD) announced that variations on CMOS transistors operate with 15-nm feature sizes. A November 26, 2001, announcement by Intel disclosed a "depleted substrate transistor" having a leakage current 100 times smaller than present transistors and, therefore, a $10^4$ smaller gate leakage power. This innovation could contribute substantially to the alleviation of heat dissipation that currently looms as a major issue. Other conventional approaches promote the use of asynchronous design or self-timed circuits operating without a single, chipwide clock speed orchestrating the tempo of each transistor. CMOS and its many variations represent opportunities for vast improvements as nanoscale dimensions are reached.

Nonetheless, there are ultimate barriers to continued CMOS scaling, and new approaches for new devices and functions are being explored. Quantum interference effects, for example, may provide opportunities for new devices and functions. Some of the new approaches are based on alternative designs for transistors, while others represent entirely new ideas for logic operations. It is clear that the current architecture for digital computers is not unique, nor does it provide the greatest capability for some operations. The brain is able to process information for operations such as image recognition with far greater speed and efficacy than current computational approaches. Just how alternative architectures may operate, and which ones are likely to provide substantial improvement for certain operations, remains a frontier of current research.

There have been many attempts to think outside the box. Radically different approaches are being investigated that, in most cases, attack only one small element of what, ultimately, must be an integrated effort tying together many factors that must be satisfied simultaneously for such a system to be of practical use (in a manner similar to the ITRS). These new approaches, some of which stem from frontier scientific discoveries, illuminate aspects of the difficult terrain ahead. Reviews of many of these approaches and observations on their performance are available.

Enthusiasm for fabricating logic circuits from molecules is currently high. Tour-de-force feats with carbon nanotubes and molecular-layer transistors have succeeded in actually fabricating transistors and even simple logic circuits. The progress made in 2001 was recognized as the breakthrough of the year by the magazine *Science*.

With such significant progress, it may appear that technological developments are imminent. However, there appears to be no single approach with a clear path to competitive products. Integration of these individual elements into a highly dense fabric with functionality approaching today's integrated circuits remains a formidable challenge, particularly in the face of our relatively rudimentary fabrication capabilities at the nanoscale. A number of approaches are being pursued with the goal of revealing additional scientific information that might improve prospects. These limitations have been discussed in articles by Meindl and Thompson, Packan and Bohr. The challenge presented by the appropriate design of interconnects has also been investigated extensively. In this context, it is important to acknowledge the extreme sophistication of the current integrated circuit paradigm and to recognize that the most likely early adoptions of these new technologies will be as adjuncts to, rather than replacements for, the manufacturing technology that currently dominates the marketplace.

A recent paper attempts to introduce some logic to the plethora of current research directions. Of great importance is the observation that the transistor, the basis for such tremendous gains in information technology, has features that are critically important to its success:

(1) it offers high gain, which allows a single transistor to reset logic levels after each stage and to drive multiple following transistors (fan-out) and (2) it isolates the output of the device from the input. These features allow the signal for a bit, with inevitably irregular amplitudes and features, to be combined with signals for other bits, yielding reliable logical functions and the accumulation of a result that maintains its integrity in a noisy environment. Other devices, such as resonant tunneling diodes (RTDs), can be used to generate gain; but because they are two terminal diode devices, they lack the isolation required for robust accumulation of logic operations. The advantages of RTDs with respect to switching speed have been known for years, but no uses for these devices have been found for logical operations in computers.

The fact that transistors have had no competitor in digital electronics for 40 years does not imply that no alternatives should be sought and studied. However, a search for a new concept must include an awareness of what digital means: a well-defined value for a digit and a way of maintaining and setting a signal to that value in a noisy environment, with mass-produced imprecise components. Alternatively it should be perfectly clear that digital representation is being abandoned if that is indeed the case, and that there is another way to cope with the inevitable uncertainty in the parameters of devices and the distortion of signals propagated in a large system.

With this caveat in mind, we studied the various technologies that are currently under investigation as possible successors to the silicon CMOS mantle.

**Single-Electron Transistors**

The operation of a single-electron device, discussed in detail in a recent review, is based on the fact that the charging energy to add an additional electron to a small island, generally through a tunneling barrier, becomes significant if the island is of nanoscale dimensions. Initial research was performed at low temperatures, where thermal fluctuations remain negligible for nanostructures in a readily achievable size range. However, fabrication of structures in the 1-nm range will allow stable room temperature operation.

Likharev points out two major unsolved challenges that face developers of single-electron transistor logic. First, there is the deleterious effect of random stray charges embedded in nearby insulators. These stray charges produce random, time-varying background charge levels, which impact device thresholds. Second, subnanometer structures will be needed at the heart of the single-electron device to allow room temperature operation. They will need to be very regular in size and shape to assure uniform device performance. If self- assembly fabrication methods are used to generate perfectly regular nanostructures, they must necessarily be incorporated into a larger microstructure. The precise placement and interconnection of these subnanometer structures, including the placement of suitable tunneling barriers, present a formidable challenge to schemes for nanoassembly. Since these devices operate at the level of individual electrons, there is a fundamental issue with gain and fan-out (the ability to drive multiple following stages from a single device output) that poses a significant architectural problem for large systems.

**Spin-Based Electronics**

Following the development of the very successful giant magnetoresistive read heads for magnetic storage and magnetic field sensors, a new area of spin-based electronics is emerging. The concept is to use the spin of the electron in suitably designed devices to perform logic operations. One can imagine that information now stored as the presence or absence of charge could alternatively be stored as spin-up or spin-down. This might be particularly attractive if the spin information can be communicated from one point in the circuit to another without moving the corresponding charges, which would eliminate the accompanying dissipation mechanisms. This research is in its initial phases, asking basic questions about the distances over which spin can be transported without depolarization, the characteristics of sources and detectors of electron polarization, the effects of interfaces between ferromagnets and semiconductors, and the feasibility of ferromagnetic semiconductors in such applications.

Domains in ferromagnets, which are formed by spins coupled together by the exchange interaction, are also being explored for use in computing. The propagation of the orientation direction of a magnetic domain through a series of nanoscale dots has been recently demonstrated. Further, using domain orientation to represent logic states, it has been possible to experimentally demonstrate the functionality of both a NOT gate and a shift register using continuous nanoscale magnetic "wires."

## MOLECULAR ELECTRONICS

In its present incarnation, the term "molecular electronics" was coined shortly after the discovery of conducting organic polymers in the late 1970s in recognition of the significant electrical conduction properties of many organic materials. A second driving factor was the clear recognition that the brain of a living species represented a logical device with the ability to recognize an image much more rapidly that digital computers. This was the proof of theorem for a "molecular computer," and the vision grew. Today molecular electronics has come to mean the use of molecules in electronic devices.

Initial visions of molecular electronics focused on how the arrangement of chemical bonds in these molecules might function as circuits and switches. A significant number of researchers are exploring this area, although even after two decades it has not been possible to experimentally verify many of the initial hypotheses. Some direct measurements of electrical conductivity across single molecules have verified the magnitude of the electrical conduction found in single-molecule layers and illuminated possible mechanisms for conductivity in bulk molecular materials. Many of the experimental observations involving molecular conductivity behaviour are puzzling and have not been explained fully.

As the behaviour of molecular units becomes better understood, it must be emphasized that a wide range of issues faces their practical application. The many considerations and challenges in the ITRS indicate the complexity of designing, fabricating, testing, and packaging chips with 0.13-micrometer feature sizes today. All of these issues are likely to be considerably more complex as dimensions are further reduced. This suggests that the first uses of molecular electronics are likely to be as adjuncts to, rather than replacements for, the integrated circuit.

As activity in molecular electronics expanded, it was recognized that molecular materials might provide significant advantages for many other electronics applications. Organic thin-film transistors (the bulk organic material) demonstrate significant gain with reasonable characteristics for some electronic devices. The mobility of charge carriers in these materials is about three orders of magnitude lower than for commonly used semiconductors. This is fundamentally due to the hopping mechanism for conduction in disordered materials (as compared with band conduction in crystalline materials, which relegates such devices to speeds much slower than those currently achieved with today's CMOS devices). But organic transistors fabricated by a variety of methods offer advantages such as low temperature, low-cost formation of large-area arrays, with particular potential for applications involving flexible structures (e.g., products such as credit cards and displays). Ink jet printers have been used to achieve transistor gate lengths of 5 micrometers and also to fabricate arrays of organic light-emitting diodes. Ink jet techniques have been also extended to such unconventional areas as deposition of suspended alloys and metallic or magnetic nanoparticles offering advantages for electronic applications. Organic transistors are envisioned for use as switching devices for active matrix flat panel displays (liquid crystal, organic light-emitting diodes, and "electronic paper"). In addition, organic and semiconductor white-light-emitting structures are anticipated to come into use in the future and to have a significant impact on energy use. All-polymer integrated circuits for use as radio frequency identification tags and for various sensors have been proposed. A variety of materials and methods are under examination with the purpose of developing low-cost, continuous-feed or reel-to-reel production methods for these low-end applications.

**Carbon Nanotube Electronics**

Carbon nanotubes (CNTs) are a unique material with remarkable electronic, mechanical, and chemical properties. CNT electrical behaviour is different from that of ordinary conductors. Depending on the application, the difference in behaviour might be an advantage, a disadvantage, or an opportunity. Electrical conduction within a perfect nanotube is ballistic, with low thermal dissipation, an advantage for computer chips if the tubes can be seamlessly interconnected. Perturbations such as electrical connections modify this behaviour substantially. For slower signal speed for analog processing, nanotubes surrounding buckyball molecules acting as transmitters have been experimentally demonstrated. Individual multiwalled CNTs at room temperature exhibit quantized conductance at values of $G = 2e^2/h$ (= 12.9 $kU^{-1}$), a remarkable observation. The current density for these experiments was $10^7$ A/cm$^2$, a value two orders of magnitude greater than current densities normally available for superconductors.

Field-effect molecular transistors have been demonstrated using a back gate with a carbon nanotube settled across two gold conductors. The challenge Because of their unique self-assembled and atomically perfect structures, carbon nanotubes exhibit unusual electrical, mechanical, and chemical properties. These special properties, such as the ability to carry exceptionally high current densities in long molecularly perfect "wires" and unusually high mechanical strength at the limit of small 'fiber' diameters, have generated much interest in the potential applications of nanotubes.

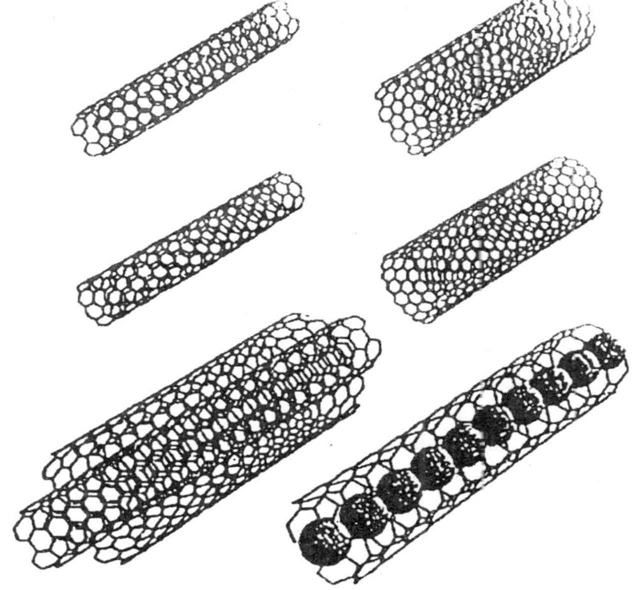

*Fig. 17.1. Carbon nanotube structures.*

### Displays

Depending on their diameter and chirality, nanotubes exhibit either metallic (like copper) or semiconducting behaviour. Metallic nanotubes can emit electrons from their extremely fine tips at quite low-voltages. The possibility of fabricating nanotube arrays on surfaces for efficient current emitting elements is of great interest for low-power field emission displays, and a number of companies are racing to develop flat panel displays for next generation television and computer screens.

### Computing

Further down the road, computer memory and logic concepts based on carbon nanotubes are being explored. Transistors made from carbon nanotubes a few nanometers in diameter—a hundred times smaller than the 130-nanometer transistor gates now found on computer chips have been demonstrated. Also demonstrated are collections of nanotube transistors working together as simple logic gates, the fundamental computer component that transforms electrical signals into meaningful ones and zeros. If nanotubes or related nanowires could be used as tiny electronic switches or transistors, computer designers could, in principle, cram billions of devices onto a chip (the Pentium 4 has only

55 million transistors). The real challenge in this or any other alternative nanotechnology approach to fabricating computer chips is to design and connect up many millions or billions of such components in a highly manufacturable and reliable architecture.

***Mechanical properties***

As a result of their seamless cylindrical structure, carbon nanotubes have low density, high stiffness, and high axial strength. Theoretical studies and recent experimental measurements suggest that the Young's modulus and breaking strength of single-wall carbon nanotubes are exceptionally high. Carbon fibers with a tensile strength up to 6 GPa are commercially available, while initial experimental measurements on 4-mm-long single-wall carbon nanotube (SWNT) "ropes" consisting of tens to hundreds of individual SWNTs have yielded values up to 45 GPa. The hope is that millimeter-long SWNTs can be formed into longer fibers or dispersed into a composite matrix while still maintaining a significant fraction of this observed improvement over conventional carbon fibers. The major challenge is retaining the strength of nanofibers and assemblies of nanofibers in conjunction with a matrix material so these properties can be controlled, optimized, and made practical.

***Energy storage***

Carbon nanotubes could be used to improve batteries. They can in principle store twice as much energy density as graphite, the form of carbon currently used as an electrode in many rechargeable lithium batteries. Conventional graphite electrodes can reversibly store one lithium ion for every six carbon atoms. Tiny straws of carbon tubes reversibly store one charged ion for every three carbon atoms, double the capacity of graphite. Carbon nanotubes are also being investigated for hydrogen storage. They may be capable of storing amounts comparable to or exceeding the U.S. Department of Energy target of 6.5 percent of their own weight in hydrogen, a level considered necessary to be practical for fuel cell electric vehicles.

The carbon nanotube is a now-classic example of a well-defined nanostructure, and exploring ways to exploit its unique properties for possible nanotechnology-based applications remains a subject of intense interest. A general issue is the ability to reproducibly obtain large quantities of selective configurations of single-or multiple-wall nanotubes. Methods for synthesizing nanotubes, controlling orientation, and producing macroscopic quantities of these nanostructures are advancing but still are at an early stage of development. Integrating these materials into, for example, a composite matrix or an interconnected electrical structure, where their nanoscale properties translate into macroscale effects, remains a key challenge.

## Impacts of Micro- and Nanotechnologies on Air Force Missions

We found four major themes that describe the impacts that micro- and nanotechnologies will have as they provide revolutionary advances for Air Force missions: (i) increased information capabilities, (ii) miniaturization of systems, (iii) new materials resulting from new science at these scales, and (iv) increased functionality and autonomy.

The dramatic increase in information capabilities of the past several decades—processing, storing, communicating, and displaying—shows no signs of imminent slowing. The integrated circuit industry is predicting $\sim 128\times$ improvements in transistor density, based on current state of the technology, over the next 15 years or so. Many other technologies are being explored for use once CMOS silicon scaling has reached saturation. This will allow the Air Force to pack more and more intelligence into smaller and smaller, lighter and lighter, less-power-consuming (per function) packages that have increasing local intelligence and increasing autonomy.

For space systems the ability to miniaturize will allow smaller and lighter vehicles, increased quality control, and new options for launch. Both space systems and avionics will benefit from the associated trend to modularization and standardization with improved reliability and better control of development costs. The adoption of the batch manufacturing and built-in-place fabrication approaches brought about by microtechnologies will increase reliability and reduce cost and will be especially enabling. The continued rapid advance of information technology will make possible new systems approaches. Miniaturization is expected to bring a world of ubiquitous sensing and computing to Air Force systems (lower power, smaller size, more function with embedded systems, lower cost, and increased capability), and the concomitant development of software for platforms, sensors, and C3I will become increasingly important.

The challenge will be to discover ways to optimize the benefit from these rapidly evolving areas of technology. Properties qualitatively change as things get small. There is great opportunity for discovering the new science and material properties that can be achieved at the micro- and nanoscale and for applying these capabilities to Air Force systems. The new properties being explored at the small scale, particularly the nanoscale, will lead to new capabilities for sensing, information processing and storage, propulsion, and high-performance materials. Stronger, more durable, and lighter-weight structures and increased efficiency in the use of energy are of particular interest. A particularly important challenge is the increased use of smart and adaptive materials—for example, to improve boundary layer control on aircraft or allow reconfigurable surveillance systems that learn from and adapt to their environment—to enable new systems approaches. Over the long term, these advances are likely to be accelerated as the study of biologically inspired systems leads to new ways to control, tailor, and adapt the properties of materials to their environment and to more efficiently store and utilize energy and information in small-scale systems.

The advances in information density, miniaturization, and materials functionality will enable an advanced degree of autonomous systems operation and a paradigm shift from reliance on a few large systems to many small things that work together. Over time, the increased intelligence of unmanned systems with integrated sensor suites and information processing (situational awareness) capabilities, and offensive/defensive reaction capabilities will probably change the character of warfare, allowing a degree of autonomy unthinkable today. The opportunities made possible by understanding and exploiting the emergent behaviour of large numbers of entities working together as a system is a long-term challenge that will lead to new system architectures and revolutionary capabilities in analogy to biological systems.

**Opportunities**

In this chapter, we explore the implications of the emerging micro and nanotechnologies for future Air Force systems applications. As discussed, four overarching themes emerge from this study of micro and nano technologies: (i) increased information capabilities, (ii) miniaturization of systems, (iii) new materials resulting from new science at these scales, and (iv) increased functionality and autonomy.

These trends are pervasive throughout the advances in micro- and nanotechnologies, and the new capabilities they provide will have far-reaching consequences for Air Force missions. We now discusses the implications of each of these trends.

**Increased Information Capabilities**

We see a continued scaling of microelectronic, magnetic, and optical devices to smaller size and higher densities at ever-higher speeds. The result will be the ability to store, process, and communicate an ever-increasing amount of information. Information, information processing, and communications networks are at the core of every military activity. Throughout history, military leaders have regarded information superiority as a key enabler of victory. However, the ongoing "information revolution" is creating not only a quantitative, but a qualitative change in the information environment that by 2020 will result in profound changes in the conduct of military operations. In fact, advances in information capabilities are proceeding so rapidly that there is a risk of outstripping our ability to capture ideas, formulate operational concepts, and develop the capacity to assess results. While the goal of achieving information superiority will not change, the nature, scope, and "rules" of the quest are changing radically.

From the analysis, it is apparent that the current rapid increase in the ability to handle information will continue at least for the next decade and beyond. The doubling of computing power every 18 months and the even more rapid increase in information transmission rate and storage capacity will lead to an increase of at least 128× in the amount of information that can be gathered and processed. Today's smart weapons will seem "mentally challenged" 10 years from now.

The trend in information density is important because so many information technologies are foreseen to have a significant impact on future military operations. Examples include these:

1. Autonomous and adaptive algorithms for resource scheduling, mission planning, and mission execution.
2. Artificial/virtual intelligence (AI/VI), self-awareness, intuitiveness, automated recognition.
3. Human-machine interfaces and robotics.
4. Heterogeneous databases, software, integration, modeling and processing techniques.
5. Advanced tools and algorithms for modeling and simulation (M&S).
6. Satellite on-board data processing and storage.
7. Nonvolatile random access memory.
8. Mass storage memory (including optical storage technologies).
9. Radiation hardening and shielding of components.
10. Plug-and-play hardware and software technologies.

Beyond this relatively near-term trend, we have anticipated that emerging nanotechnologies will enable even more revolutionary long-term changes in how we obtain and use information. Exploiting these advances will be an important and challenging task for the Air Force.

**Air Force Missions**

Micro- and nanotechnology's potential for reducing weight and size while enhancing performance has particular relevance to the Air Force mission of defending the United States through control and exploitation of air and space. The benefits of significant miniaturization, reduced cost, and increasing performance, if they can be achieved, would be particularly significant. The advance of information technologies provides the clearest demonstration of this promise (one place where it was especially influential is avionics). Emerging areas such as MEMS and integrated sensor systems point the way to new opportunities. Miniaturization of systems, combined with batch fabrication and integration of components, may lead to significant improvements in affordability, enabling more densely distributed systems. Further, improved performance of materials through nanostructuring and other advances in nanoscience and technology could enable systems opportunities of a more revolutionary nature, so that such advances merit careful tracking by the Air Force.

Future opportunities of micro- and nanotechnologies are relevant to all six of the core competencies in the Air Force strategic plan: (i) aerospace superiority, (ii) information superiority, (iii) global attack, (iv) precision engagement, (v) rapid global mobility, and (vi) agile combat support.

Increased functionality in smaller packages will enable greater reach, more precise engagement, greater global awareness and knowledge, and more agility in the control of air and space. Understanding how the capabilities of advancing micro- and nanotechnologies can be exploited in new systems that will enhance missions is a major challenge worthy of significant attention when planning. Changes are likely to be required not only in specific platforms but also in the organization of the Air Force to fully exploit the game—changing advances that are likely to flow from the application of micro- and nanotechnologies to its missions. These are well beyond the scope of the present report but should not be ignored as these technologies advance into systems and platforms.

The Air Force will be challenged to use the advances in micro- and nanotechnologies in its hardware and systems. There is a long and wide gap between the discovery of a technology and its commercialization to the point where it can be incorporated into military hardware. Coordination of S&T investments while considering both Air Force mission planning and external advances in science, technology, and commercial development will be critical. Although it is impossible to predict with certainty the systems implications, mission impact, and commercial viability of these emerging advances, it is important to consider their possible benefits. The continually changing competitive environment with respect to threats from adversaries underlines the need for alternative approaches to some of the most daunting threats. Based on the technologies discussed, we now discusses some specific opportunities that could emerge from Air Force investments in micro- and nanotechnology S&T in these areas.

## Areas of Opportunity

In this section we have identified selected opportunities for linking micro- and nanotechnologies to Air Force mission platforms. While pursuit of any of the specific systems opportunities depends on considerations well beyond the scope of the present study, it would be beneficial to examine these or related opportunities to determine their fit with Air Force needs. As appropriate, key issues should be further examined through focused S&T projects within the AFRL, support to the broader R&D community through focused AFOSR grants and AF contracts, and related program strategies.

An overview of several opportunity areas is given in which couples the opportunities to the micro- and nanotechnology S&T areas as discussed (left side). Also shown is the relationship of these opportunities to the Air Force core competencies (right side). The approximate time frame in which these technologies could have an impact on Air Force missions is near term (0-10 years), medium term (10-20 years), or long term (20-50 years). The technologies are by use: space vehicles and systems, weapon systems, and air vehicles and systems.

### Space Vehicles and Systems

The miniaturization of systems enabled by micro- and nanotechnologies while maintaining or increasing capabilities provides an important opportunity for space systems. These advances will enable satellites to carry out more functions consistent with increased information superiority and to employ lighter vehicles or arrays of satellites, providing new options for launch vehicles and for distributed sensing or communications from space. Because of their reduced weight and cost, nano- or picosatellites may enable new functions.

At the same time, adopting the manufacturing advances pioneered in the microelectronics industry—batch processing and increased quality control—along with anticipated future advances in nanotechnologies such as self-repair and reconfigurability will significantly expand the range of possible applications. Low-power systems in combination with more efficient space power and energy storage systems will enable long lifetimes. Future ultrahigh-resolution sensor systems will be enabled by microarray technology for imaging sensors, on-board digital image processing, and wide band communications. Large arrays, made possible by miniaturization and the associated reduced weight and cost, and phased arrays will enable continuous surveillance. The ability to monitor with high spatial resolution over essentially all wavelengths will allow truly realizing continuous total information systems. Specific areas suggested for further consideration are distributed satellites, integrated spacecraft, and micro-launch vehicles.

***Satellites***

Two or more satellites with suitable coherent signal combining can have the resolving power of a much larger spacecraft. The angular resolving capability is roughly equal to $e/D$, where $e$ is wavelength (about 0.5 micrometers for visible light and anywhere from millimeters to a meter for radio frequencies) and $D$ is the separation between spacecraft. Distributed spacecraft are the only practical way to generate milliradian-wide beam widths in space at UHF frequencies (300-1,000 MHz); the required kilometer-scale antenna diameters are impossible using a single antenna structure. Possible applications include battlefield cellular communications systems, where the cells are generated by extremely narrow spot beams from satellites, and geolocation of UHF jammers. Geolocation can be accomplished using two spacecraft, while synthesis of a narrow beam antenna for communications will require hundreds of individual spacecraft swarms in a sparse aperture array. Mass-produced micro-, nano-, and picosatellites with orbit and attitude control capability make the latter scenario possible. The key technology development that is still needed for such arrays to work is handling the phase of the signal from this multitude of separate transmitters and receivers. Technology developments in low-power, space-hardened digital and radio-frequency electronics, microthrusters, and low-mass, low-power, short-range (kilometer-scale), free-space optical transmission systems are required. Collective arrays of satellites that function in a synchronized fashion promise significant new opportunities in capabilities and robustness of satellite systems.

**Integrated Spacecraft**

Semiconductor batch-fabrication techniques allow cost- effective production of million-transistor digital circuits, analog circuits, radio-frequency and microwave circuits, and microelectromechanical systems (MEMS). Current trends in semiconductor and MEMS fabrication, e.g., continually increasing functionality per square millimeter of silicon, argue for the development of system-on-a-chip (SOC) technology for various spacecraft systems. Increased electronic functionality, coupled with nonvolatile memory technologies such as FLASHRAM and MRAM, enable intelligent micro- and nanosystems that can be reprogrammed or reconfigured on orbit. Possible examples of spacecraft SOCs include Sun and horizon sensors, inertial measurement units (IMUs) composed of MEMS accelerometers and rate gyros, GPS receivers for navigation and attitude determination, and MEMS-based microthruster systems. SOC interconnection can use high-speed serial lines to allow plug-and-play assembly much like the USB bus used for personal computer peripherals. The challenge will be to migrate commercial technologies into radiation-hard or radiation-tolerant technologies for use in space. The benefit will be decreased parts count per spacecraft, increased functionality per unit spacecraft: mass, and the ability to mass produce micro-, nano-, and picosatellites for launch-on-demand tactical applications (e.g., inspector spacecraft) and distributed space systems.

**Micro Launch Vehicles**

Enabled by subsystems and components such as MEMS liquid rocket engines, valves, gyros, and accelerometers, microlaunch vehicles are feasible in the size range 15 to 800 kilograms gross liftoff weight (GLOW). The payloads would make extensive use of micro- and nanotechnology as well. As an example, a 170-pound GLOW, two-stage rocket (the weight of an AIM-9) could deliver one or two kilograms to low Earth orbit. Such vehicles would be about the size and complexity (less the seeker) of a small tactical missile and thus should cost about the same. Launched from the ground or the air, the micro launch vehicle redefines the concept of low-cost access to space as cost per mission rather than cost per pound of payload (which is about the same as for larger launchers). Thus, it will be possible to place a payload (albeit a small one) into orbit for $10,000 to $50,000 rather than the $10 million to $50 million required today. This cost is sufficiently low that launchers can be stockpiled for launch- on-demand access to space, encouraging routine tactical space operations. Air launch (from a tactical aircraft for example) offers the added advantages of launch site flexibility, elimination of the requirement for a launch range with its attendant costs, and covertness (a liquid-fuelled vehicle this small might be very difficult to observe from space; also, there need be no fixed launch location). Orbital applications might include visual and IR inspection of space objects, ELINT, jamming of satellites, and antisatellite operations.

## Weapon Systems

### *Miniaturized ballistic missiles*

The same MEMS technologies (propulsion, guidance, control, etc.) required for microlaunch vehicles enable intercontinental tactical ballistic missiles with sensor or nonnuclear munitions payloads. A 170-pound GLOW, two- stage rocket (about the weight of an AIM-9) could deliver about 10 pounds to 4,500 nautical miles or 30 pounds to 1,000 nautical miles (the reentry vehicle might account for 25 to 40 percent of this weight). The advantages of the ballistic approach compared with the air-breathing cruise missiles approach are very high speed and immunity to interception. Global, rapid deployment of sensors (and even micro air vehicles) is one obvious application. Another is weapons delivery. Enhanced effectiveness warheads pack sufficient lethality into a small package that these small missiles may be effective against relatively soft targets such as radar sites and armored vehicles. Now that much of the target detection and acquisition is done off-board the shooter platform, the long range and high speed suggest that this may be an effective approach to the now-difficult problem of attacking rapidly redeployable or moving targets, such as missile launchers in Iraq. The ultralong range eliminates the "tyranny of distance" that plagues operations in places like Afghanistan, since any target in the world is within practical range. This is truly a global reach system.

### *Missile boost phase interceptors*

Air-launched boost-phase intercept schemes could greatly increase the payload and endurance capabilities of launch platforms. Ground-launched missile interceptors are now sized according to the sensor and propulsion systems, since a direct-hit, kinetic-kill vehicle need not carry a warhead. Advanced sensors, propulsion, and other subsystems enabled by micro- and nanotechnologies will permit dramatically reduced missile mass (perhaps well below 100 kilograms), with several profound system-level impacts. One impact is that microinterceptors will have unit costs at least one and possibly two orders of magnitude lower than current designs. The miniaturization could enable UAV-launched missile boost-phase interceptors. Multiple, simultaneous launches against a single target could increase reliability, reduce the performance requirement relative to that needed for target-decoy discrimination, and permit new, more flexible system architectures. This may, for the first time, shift the economics of missile offense versus defense in favour of the defender.

### *Air-to-air and air-to-ground weapons*

Advances in micro- and nanoscale technologies offer the opportunity to explore designs for air-to-air missiles and air-to-ground bombs with significantly better performance. It may be possible to reduce weight, size, and cost by miniaturizing of sensors, avionics, and inertial measurement units in both the near and the far term. Recent improvements in long-range missile accuracy have resulted in significantly reduced loss of expensive missile-carrying aircraft in combat situations. Adoption of the next generations of MEMS-based IMUs, providing further improvements in missile reliability and accuracy, will be accelerated to further improve aircraft combat survivability. Reductions in guidance system cost and package size will translate into system savings for short- range air-to-ground and air-to-air missiles. Coupled with enhanced explosives under development and further evolutionary cost and accuracy improvements in MEMS IMUs and seekers, further reductions in size and cost may be realized, allowing significantly larger numbers of missiles or smart bombs to be carried by an aircraft. This payload quantity extension can be traded for range or logistics support size and cost as overall systems requirements dictate.

## Air Vehicles and Systems

### *Micro air vehicles*

The acronym MAV generally refers to 6-inch or smaller flying platforms, which are under development for local surveillance and reconnaissance applications. Given micro- or nano-enabled or -enhanced sensors and subsystem technologies, MAVs can grow dramatically in capability, achieving autonomy and the functionality of systems that are currently many times larger, or can shrink in size to insectlike dimensions. Current MAVs fly at 10 meters per second for 5 kilometers, but future systems could fly transonically for 1,000 kilometers or endure for tens of hours.

In addition to performing sensing, surveillance, and reconnaissance, enhanced capability systems might function as microdecoys or carry jammers (which need not radiate much power if they are very close to the receiving antenna) or even be weaponized. Replacing larger air vehicles with MAVs promises to greatly reduce cost. The affordability of MAVs could allow large swarms of vehicles with cooperative behaviour and could herald new and more robust systems approaches to reconnaissance and surveillance. Also, the lower overall mass of a system-level solution implies a dramatically reduced logistics tail. Low cost might be traded for area coverage (by using more vehicles) or operation in very high risk environments. Large numbers flying under the tree canopy or perching in cities may be one solution to finding hidden targets.

***MEMS aerodynamic flight control***

Microsensing and control of airflow over vehicle surfaces, combined with new ultrastrong, lightweight materials, could lead to a new generation of aircraft providing enhanced flight efficiency and maneuverability without conventional rudders or other macroscopic control surfaces. These MEMS-based active aerodynamic flight control vehicles (MACs) could exploit advances in microscale sensors and actuators in combination with information technologies to provide local feedback control. Vehicle surfaces would rapidly sense and change airflow boundary layer conditions, a capability now possible with micromechanical devices and continuously increasing computing power. Such control strategies might reduce, on average, the turbulent nature of aerodynamic flow, leading to laminar flow vehicles with dramatically greater range- payload capabilities than those of current aircraft. The implications for air combat support, global reach, and reduced overseas footprint could be significant. A second aspect would be the ability to manipulate boundary layers to generate large forces and moments for flight control, possibly supplementing or replacing large-scale control surfaces while reducing weight and increasing maneuverability. The possible mission implications of MACs would appear to be worth exploring in concert with the advancement of micro- and nano-technologies.

# 18

# CHANGING BATTLESPACE

America possesses the strongest military in the world. In terms of manpower equipment training, and capabilities its armed forces have distinguished America as perhaps the only remaining superpower. This military might has been crucial to America's prosperity—the possession of great wealth requires the capability to defend it. It has also been pivotal to the geopolitical stability of the world. After the invention of nuclear weapons in 1945 and the Cold War that followed it, America, Russia, and the other nuclear powers developed large, robust arsenals that could not be eliminated with a single preemptive strike. Any attack on one of these nations would, in principle, lead to nuclear retaliation and to the total devastation of all the parties involved. This military doctrine, sometimes called Mutually Assured Destruction (or ironically, MAD), dominated strategic thinking for much of the Cold War. It also meant a shift in the methods and locations of war since another world war is unthinkable to sane leaders in the context of MAD. Instead, as JFK pointed out, the theaters of war changed to Asia, the Middle East, Africa, and other less-developed areas.

With the collapse of the Soviet Union and the end of the Cold War came a corresponding change in the threats to America. Most Americans (and citizens of other developed countries) are no longer concerned about invasion, as we were until World War II, or about extinction, as we were during the Cold War. In this era of globalization, we are concerned about terrorist strikes that result in economic or political instability as well as death. It's useful to look at why this shift occurred and at the change in thinking that has accompanied it in order to understand the new challenge to America's armed forces.

While this discussion is necessarily simplified and highly abbreviated, it still adds context to the issues.

During the Cold War, America and the Soviet Union were reluctant to fight or even skirmish with each other directly since matters might escalate out of control. Instead they tended to arm and train local militias, insurgents, and armies in the nations they considered important to their national interests, such as Vietnam, Afghanistan, Cuba, Korea, Iran, and Iraq. Indigenous troops were taught how to use modern weapons systems and how to fight guerrilla wars. The super powers also tended to back governments of convenience, generally defined as those that would be amenable to the superpower's own political or economic interests rather than those that were legitimately elected. This led to ongoing strife and a culture of violence that has caused many of these nations to remain hot spots to this day. While terrorism predates the Cold War, the overwhelming majority of today's most notorious terrorists, including Osamabin Laden, either come from hot spot countries or were trained there. These people are under no illusions that they are capable of fighting the United States directly to achieve their objectives. Instead, they are trained in unconventional war and, like good martial artists, apply force precisely at their enemy's weakest points often using his own strength against him. This means killing civilians, toppling symbols and monuments, disrupting the economy, and where possible using the other side's own technology.

During the world wars and all preceding conflicts, victory was paramount. Domestic casualties were minimized when practical but a fairly high casualty rate was considered acceptable. Enemy casualties, both military and domestic, were hardly given notice. This was the psychology that led to the firebombing of Dresden and to the nuclear destruction of Hiroshima and Nagasaki. In the Korean War, and to a much greater extent in Vietnam, this began to change. Americans questioned why troops were being sent for dubious purposes of a faraway nation that most people had never heard of. The human losses became unacceptable. Enemy casualties have also now become a major issue when fighting in countries that sponsor terrorism. This is attributable to a few causes. Most such countries are not representative governments and many feel that the citizens should not suffer for the sins of their leaders. Also, some feel the pangs of a guilty conscience, for the United States helped to create many of the situations that have led to the surge in world terrorism. While very few question that a country is justified in striking back against an attacker, the debate over the recent war in Iraq (especially Germany's position) has underscored the philosophical division on the policy of risking additional civilian deaths in an effort to stop terrorism.

While these changes are certainly profound perhaps the greatest changing influence on the face of war is not directly geographic political or psychological but economic. The spread of free markets as well as advanced technology such as transportation systems, computer networks, and telecommunications systems have resulted in an economy that grows more and more globalized. In 2003, Germany, France, and Russia all had financial interests in Iraq, which made their opposition to the war seem as much like economic pragmatism as true idealism. American policy was equally stigmatized by accusations that oil needs rather than security concerns were behind its actions. More generally, the economies of the developed world are linked so closely that a ripple in any one affects the others. Pain in South America has echoes in Russia and in Europe. A recession in America or Japan brings down the rest of the world. Free-trade regions that lower the barriers still more are cropping up in Europe, the Americas, Asia, and possibly even Africa.

America has the strongest economy as well as the strongest military on earth, so it must play its cards carefully. It is clear that war and instability are counterproductive to the global economy. Threats of conflict depress consumer spending, travel, international trade, investment, and stock prices, leading to bankruptcies, negative growth, and unemployment. These issues are of surpassing importance to America, which is a large part of why the first President Bush failed to win reelection even after prosecuting a successful campaign in the Gulf War. This has led military thinkers such as General John Sheehan to speculate that the core mission of the U.S. military is now not just to protect America, but to help maintain global stability for economic development. Many military leaders, including General Colin Powell, have become more dove-ish than their political counterparts, and it has already been largely forgotten that the cabinet post now called Secretary of Defense was not long ago called Secretary of War. Unlike in the days of empire, war no longer pays.

These elements form the challenge to American military, if at all possible conditions leading to war, such as the accumulation of weapons of mass destruction by rogue states or those that sponsor terrorism, must be detected and stopped before they become problems. When wars do happen, they will be fought in faraway places and harsh environments such as swamps, jungle and deserts. Neighbouring countries may not he friendly, so operations may need to be headquartered far horn the front lines. American and civilian casualties are unacceptable. Enemies are skilled guerrilla fighters who are relatively well armed, don always wear uniforms or move in units, and have made extensive use of unconventional weapons including chemical and biological ones. These enemies also have very few constraints: they are willing to use any weapon and any tactic necessary and they seldom worry about casualties on either side. Finally, the wars must be short to minimize economic consequences but peacekeepers may need to remain in the fields for years after a conflict.

This was, more or less, the scenario of the 2003 war in Iraq. Add to it the continuous press coverage that largely eliminates cheating or secrecy, and the Iraq mission seemed impossible even with DoD's half a trillion dollar budget. To its credit, the military stepped up to the plate and embraced the challenge. It made many changes to its organization such as relying more on intelligence and special forces groups, improving communications at every level, unifying command 0 domestic defense under the Department of Homeland Security, and increasing its capabilities for working in small, highly coordinated units. These changes are beyond the scope of this book, but they are an important first step.

The military was also very lucky. Despite concerns, no weapons of mass destruction were used—even found—during the offensive in Iraq. If they had been, the situation might have turned out quite differently. Yet this is far from the only example of how vulnerable even the strongest military in the world can be in the face of its new challenges. The solutions to many of these problems are technological, not organizational. The core research operations of the Department of Defense, including the Defense Advanced Research Projects Agency (DARPA) and the Office of Naval Research, have issued lists of Grand Challenges. These are the crucial technological hurdles that must be overcome for the military to fulfill its new mission. Nanotechnology is key to the realization of almost all of these challenges, and the rest of this chapter is dedicated to describing many of them and explaining how they are being addressed.

## Agents of Mass Destruction

The use of chemical and biological agents as weapons of mass destruction is not a new phenomenon. People have been using biological weapons ever since they started throwing decomposing corpses and feces over city walls during sieges in hopes of causing the plague. Some of the first chemical weapons were poisons used on arrowheads. Chemical and biological weapons date back to prehistory.

Despite their early use, chemical and biological weapons never gained much popularity until modern times. Chemical and biological agents are generally difficult to "weaponize" or make into devices suitable for use in combat. Before missiles and bombs were common, this was a huge problem, since a plague represented a threat to attackers as well as to defenders when troops were closely mingled. In addition, lingering contamination would remain a problem for citizens and occupiers after the battle was over. Also, a change in wind or weather could blow a toxic cloud back at an attacker, and a storm or rain shower might reduce or negate its effects. Without a mechanism for delivering them to an opponent's homeland, chemical and biological weapons have virtually no value as weapons since they are so volatile and unpredictable. It is hard enough to protect an attacking army from such weapons. Until now, it has been virtually impossible to protect a civilian population but nanotechnology may change this.

World War I, which saw the first large-scale use of aircraft and modern field artillery, was also the battlefield on which chemical and biological weapons truly reached maturity. Blister agents (also called mustard gas), nerve gas, and a litany of other horrors killed troops and civilians in unprecedented numbers. Today, toxins such as ricin and sarin have been developed to be deadly in quantities of much less than a gram. To put this in perspective ricin is considered twice as toxic as cobra venom. These agents are ideal weapons for terrorists since none of the factors that make them poor military weapons apply to terrorist tactics. Terrorists do not consider the unpredictability of these weapons to be detrimental since their psychological effect is undiminished. They are not concerned about weaponizing them since they can easily use suicide bombers to attack people or to plant toxic compounds directly into a building's ventilation system or a city's water reservoir. Sarin was used in this way when the Aum Shinrikiyo Supreme Truth cult distributed it on the Tokyo subway in 1995, killing a dozen people and injuring thousands. Ricin also has a history—it was a ricin pellet in an umbrella tip that was used in the 1978 assassination of Georgi Markov, the Bulgarian writer and journalist.

*Chemical weapons* are easy to make from common materials. Ricin, for example, is derived from castor beans with the aid of some basic chemistry apparatus, all of which are available in any high

**Table 18.1. The relative toxicity of some chemical weapons and biotoxins.**

| *Agent* | *Lethal dosage (LD 50 g/kg body weight)* | *Source* |
|---|---|---|
| Botulinum toxin | 0.001 | Bacterium |
| Diphtheria toxin | 0.10 | Bacterium |
| Ricin | 3.0 | Castor bean |
| VX | 15.0 | Chemical agent |
| GB | 100.0 | Chemical agent |
| Anthrax | 0.004-0.02 | Bacterium |
| Plutonium | 1 | Nuclear fuel |

school science laboratory. The British authorities have already caught suspects trying to make the substance in London and there is some evidence that it was recently tested in northern Iraq. Hydrogen cyanide (HCN) is another easily obtained chemical agent. It is mass produced for industrial uses such as tempering steel, dyeing, engraving, gold mining, and the production of some plastics. It is also the active ingredient in Zyklon B, the gas used in Nazi gas chambers. It is still used in parts of the United States for criminal executions.

Unlike explosives, almost all of which are based on materials with a chemical similarity (they contain nitrates), there is no single commonality among all chemical or biological weapons. Even nuclear weapons are fairly easy to detect via the radiation they emit. Radiological contamination can be accurately and quickly measured using a hand-held Geiger counter, but there is no similar device for chemical or biological weapons. This makes detecting them by any current technique extremely difficult, since most equipment relies on automating a single test. "*Bomb sniffers*" and the detection equipment used at airports will catch most common types of explosives by testing for nitrates, but most chemical and biological weapons will get right through.

Today, the best approach for determining if a chemical weapon is present involves a technique called *spectroscopy*. Spectroscopy uses a sample's absorption scattering, and other interactions with particular frequencies of light to develop a chemical fingerprint. This is the same approach that astronomers use to determine the atmospheric composition of planets and stars. Unfortunately, it can require a reasonable amount of laboratory equipment and a good quantity of a sample (or a large-scale telescope in the case of stellar measurements).

Results can sometimes take hours to obtain. Biological agents are no easier to detect. When *anthrax-tainted letters* were sent around the country in 2002, it took three full days to give the all-clear at most suspected sites. This kind of time lapse is unacceptable for troops in the field. If an area may be contaminated or a shell may contain chemical or biological agents soldiers on the front lines will have been exposed before they can react, and troops coming up behind them have only minutes to prepare.

In Iraq, the army used pigeons and chickens as an early *chemical weapon* warning system in what some servicemen jokingly called Operation KFC (for *Kuwaiti Fried Chicken*). The idea is that these birds are more vulnerable to the effects of poison gas than people are. If they die, soldiers can conclude that there is a poison in the atmosphere. This was also the reason that miners took canaries into coal mines in the 19th Century. While better than nothing the problems with this scheme are manifold. For a start, almost anything from blowing sand to heatstroke can kill a chicken, so there will be plenty of false alarms. More importantly when much less than a gram of a toxin is enough to kill a persona by the time the chicken dies it is already too late.

Even assuming it were possible to detect chemical and biological toxins instantly and accurately, how could troops be protected from them? Currently, it is necessary to keep troops inside protected vehicles or to make them wear heavy protective Suits that cover the entire body and prevent anything from getting in. This is similar to trying to fight in a space suit, and it isn't practical for active combat or for any mission requiring more than a few hours exposure in desert or jungle conditions.

After the dust settles and the toxins have been detected, what options are there for destroying them arid decontaminating an area? If the toxin is contained within a shell it is relatively easy to destroy, but once it is distributed, it is much more difficult. Some equipment can be power-scrubbed or -cleaned, but the process is expensive, time consuming, and results in a great deal of contaminated water. What if the contamination is in the desert where water is scarce? What if the water supply itself is contaminated? While some biological toxins can be neutralized by boiling or removed by micro-filtration, it is often hard to explain this process to local people in hostile territory, and many chemicals cannot be removed this way. Once they are infected, people can also transmit anthrax and other diseases used as weapons.

How can people be tested quickly and quarantined before they infect others?

These scenarios are more likely than they seem. Saddam Hussein used chemical weapons on Kurds in his own country. He also used eco-terrorist tactics in the Gulf War by setting fire to oil wells. He might well have used chemical weapons to slash and burn his territory had he remained in power long enough after the invasion began. It is clear that if chemical or biological weapons had come into play in Iraq or do come into play before adequate countermeasures are developed, shock and awe military tactics will not help make the U.S. military any less vulnerable. Something much more defensive and more powerful will be needed.

## Nanosphere

Enter nanotechnology. What if the desk-sized labs that could determine the presence of a chemical or biological toxin could be integrated onto a chip the size of a fingernail? What if they could also perform their tests in seconds or even fractions of a second for the more common toxins? This would make them lightweight enough for any soldier to carry, and quick enough to allow that soldier to take action before it is too late. This may be possible using a nano technology technique called *lab-on-a-chip*. Labs-on-chips are complete chemical or biological laboratories including chemical reactor cells, sample storage, and micro- or nanoscale channels, all integrated onto a microchip in much the same way that circuits are.

Companies such as *Nanosphere* and *Microsensor systems* have already developed the first few devices utilizing this approach for biotoxin detection. Right now these sensors are still the size of laptop computers but they are getting smaller rapidly. There is reason to believe that thumbnail-size sensors will be possible since companies including Agilent (formerly part of Hewlett Packard) and Affymetrix have already made such small sensors for simpler kinds of analysis

A sample device from *nanosphere*, a company specializing in DNA-based sensor nanotechnology. The sensor contains nanoscale particles of gold (often called nano dots), and each of the gold nano dots has several short strands of specifically designed DNA attached to it. The DNA attached to the nano dots will chemically bind to a certain target DNA and to no other; for example, it might bind only to the DNA anthrax. If anthrax is present the DNA strands connected to the nano dots will bind to it, drawing the nanodots closer together. As they draw closer together, the *nano dots* undergo a change that only happens at the nanoscale—they change colour without changing structure or composition.

The reason for the colour change is complex and has to do with the coupling of the properties of quantum mechanics with the properties of the visible world, but the result is straightforward. In the presence of anthrax (or whatever *biotoxin* the sensor is designed to detect) and under no other conditions, the sensor will change colour—in this case, from red to blue. The nano scale colour change may be detected by a laser and transmitted to a soldier via a computer or heads-up display. Alternatively, if a more direct reading is desired, the soldier could carry a visible quantity of these sensors, effectively a litmus paper for biotoxins. While the science behind it may be complicated, this is a test that anyone can use.

This technology and several other promising candidates are evolving quickly, and soon all that will be needed to develop a test for a new toxin is the toxin's genetic fingerprint. The same technology can also be used for diagnosing diseases. In the case of SARS, a genetic fingerprint was found within three months. As DNA sequencing technology continues to develop, this discovery period will get shorter and shorter. With a library of biotoxin fingerprints, a sensor for each toxin could be integrated into a lab-on-a-chip and a universal, nearly instantaneous, very sensitive, and essentially 100% accurate biotoxin sensor could be a reality.

Once a sensor the size of (or very much smaller than) a thumbnail is developed, it could be installed anywhere on a soldier's uniform. Some toxins are very volatile (they evaporate easily) and some are very stable. If a sensor is located on a helmet, for example, it would not be very helpful in detecting *botulinus toxin* in a puddle, whereas a sensor in a boot might not be as helpful in detecting airborne toxins. Ideally, dozens or even hundreds of sensors might be woven into a uniform and interconnected by tiny wires running throughout the fabric. This kind of matrix could still remain extremely light weight and could also serve as a general communications mesh for all of the soldier's equipment, allowing many applications that we will discuss later. A primitive version of this kind of wire mesh is currently used in fencers' jackets to detect the impact of an opponent's sword during competitions.

An alternative to weaving sensors into a uniform is to move the sensors around so they are exposed to everything a soldier is exposed to. One plan for doing this involves injecting sensors into a soldier's bloodstream. The sensors would circulate through the bloodstream and could be monitored at a place where blood vessels are closest to the surface, such as in the eye. The monitoring unit could be installed in goggles or on a microphone boom of a communications headset. While quite invasive, these so-called in vivo sensors could also have other uses in continuously monitoring the health of a soldier.

Since soldiers on the battlefield are already linked by radio communications, any sensor solution should also be tied into the link, perhaps on a sèparate data channel. If each sensor on every soldièr could intercommunicate, a whole unit could start reacting immediately to the presence of a threat as soon as a single sensor was triggered. Also, statistical analysis could help weed out any false positives, making a unit's awareness of threats as close to perfect as possible.

**Artificial Materials**

Once a *toxin* has been recognized, the next thing to do is to protect the soldier. Since some types of nanotechnology-based sensors can be accurate to the level of detecting individual molecules 0 a toxin, there should be some advance warning before enough toxin to cause fatality has been delivered. How ever, the warning could come only seconds before a lethal dose, far too short a time to put on a chemical defense suit—especially in the middle of a firefight. Fortunately, nanotechnology provides an answer for this, too.

Consider using nanotechnology-enabled smart materials for other purposes. Materials already exist that change their properties when exposed to different external stimuli. Human skin is a good example. It tans in the presence of direct sunlight, sweats when it gets too hot, develops calluses where it is frequently abraded, and heals itself when it is punctured. For these reasons, nanotechnology is of interest to Estee Lauder, Revlon, and others whose customers don't usually dress in combat fatigues. *Artificial materials* with skin like properties have been created, but it is necessary to engineer at the molecular level to create what is required for chemical defense. The idea is to create a fabric that is a comfortable open-weave combat uniform under most conditions but, when exposed to a certain stimulus (an electrical signal from a sensor, for example), could restructure itself at the nanoscale to become an air tight shell. Air could be delivered via an oxygen re-breather of the type used by some divers or through nanotechnology improved gas masks.

Such a uniform is not only possible, but likely. Nanotechnology-enhanced fabrics that are totally resistant to the penetration of liquids (and are thus stain resistant) are already made by companies such as *Nano-Tex*, whose fabric is used in khaki pants sold by Eddie Bauer and others. While you might expect rubber or nylon pants to reject water and stains in this way (although why you would wear rubber pants is another question), only nanotechnology can combine the properties of comfortable cotton fibers with the desirable properties of repelling liquid to make fabric that is indistinguishable from ordinary khakis until you spill on it. Employ so-called *smart material* in soldiers' uniforms is just an evolution of the techniques already in use. As we'll see later, that is only one of the ways nanotechnology will affect soldiers' uniforms.

It is important to protect a soldier in his uniform, but better defenses are also required for buildings, vehicles, and other installations. Filtering out biotoxins is in some ways simpler than filtering out chemical weapons, since a single virus or bacterium is more than 100 times larger than a small molecule such as HCN or oxygen. This means that a filter with pores measuring tens of nanometers can be made to block out most biotoxins but to let air through. Unfortunately, these filters don't work well for chemical weapons where the difference in size between air and toxin may be very small indeed.

Right now, most chemical weapons protection is based on a similar technology to World I—*era gas masks*. A simple air filter is impregnated with activated carbon (also used in commercially available water filters) and other chemicals that adsorb many common contaminants. This approach is far from effective in many cases. For example, HCN is one of many common chemical weapons that does not react much with activated carbon.

A different approach involves making an artificial membrane and permeating it with nanotubes. *Nanotubes* are shaped like drinking straws made of carbon, and their diameters can be controlled with sub-nanometer precision. Using this approach could create a molecular sieve that would block all but the very smallest molecules—oxygen and nitrogen, for example and make a very good filter indeed. This effect depends on tailoring the sizes of individual molecules and is only possible through nanotechnology.

**Cleaning the Contamination**

If an area is contaminated with chemical or biological agents, it must be cleaned up in a process commonly called *remediation*. Remediation can be done in a variety of ways, including old-fashioned soaking with soap and water, forcing breakdown via chemical reaction, or exposing to extreme heat. Nano science offers a few ways to vastly improve the efficiency of this process. Consider the remediation approach of breaking down a toxic substance through a chemical reaction. A material's surface area and its reactivity are directly related. The greater the surface area is, the greater is the exposure to other materials with which to bond and interact. This is one way in which catalysts (materials that increase the speed of chemical reactions) work. Grinding a substance finer and finer increases surface area and the nanoscale is the logical limit. A quarter ounce of nano particles may have a surface area as great as a football field.

Some powders, such as *aluminum oxide* and *magnesium oxide*, have proven effective in neutralizing chemical agents and are already in use by the army. Reducing these particles to the nano scale not only increases their effectiveness by increasing reactivity, but also allows them to be impregnated into the fibers of filters like the gas mask filters we just discussed. Effectively this allows them to do what activated carbon did in older filters—aggressively bind to and remove chemical agents. While activated carbon does work this way, it is only one example of a whole family of nano particles that can be developed to deal with a host of chemical weapons. Activated carbon was created by trial and error rather than through a knowledge of the underlying mechanism that makes it function. However, research in nano science is revealing these mechanisms and is allowing the design of new nano particles specifically engineered to react with many chemicals that activated carbon can't handle.

The examples above are some of the ways in which nanotechnology will protect soldiers from *chemical* and *biological weapons*. Sensors will instantly detect the threat, transmit it to a central control unit, and initiate countermeasures. The soldier's uniform will immediately start to defend against attack by becoming nonporous and filtering the air supply. *Nanoparticles* can then be sprayed, released, or even delivered via artillery shell or other munitions to start breaking down the *chemical* or *biological agent*. Clearly these technologies have other dual-use applications including medical diagnosis using DNA-based sensors and water purification using remediation nano particles. There are also a number of other nanotechnologies that may be effective in countering weapons of mass destruction, but this is a great beginning.

## Battlefield Computers

In ancient Greece, a soldier went into battle carrying over 70 pounds of weapons and equipment, including heavy bronze armor, spears, a shield, and a sword. A modern soldier carries at least as much, though the nature of what he carries has changed significantly. Today's soldiers are expected to carry not just weapons and body armor, but radio equipment, optics such as night-vision or telescopic scopes, *global positioning systems* (GPSs), *battlefield computers*, food, water, spare clothing, and much, much more—the batteries alone can weigh more than 20 pounds. These tools are essential for net worked warfare and battle space awareness. It is crucial for troops to gather information, coordinate, maneuver, respond to enemy action, and remain in touch with commanders even in the heat of battle.

While these tools are effective, they also have serious problems. Seventy to ninety pounds of equipment are unwieldy and lead to fatigue. The systems are not integrated with each other—most hand-held range finders cannot automatically transmit information to weapons systems, and binoculars and scopes don't transmit image information, Geiger counters don't directly relay danger information to other units, and most devices can't even share power supplies so they require heavy battery units or other portable power supplies. While many devices provide a soldier with information, the devices do not process it or respond to it automatically. *Ballistic defensive* technologies (such as Kevlar vests) offer no protection for the arms, legs, or head, and they don't monitor vital signs or take action if a soldier is wounded. In many respects modern soldiers are equipped much like Spartan warriors, but with a better spear, tougher armor, and a lot of electronics. This bundle of equipment has caused them to be compared to a Christmas tree with ornaments hanging from it. The green uniform probably does not help either.

The soldier of the future could look very different. A great deal of effort is going into making soldiers more efficient and increasing survivability, a soldier's ability to avoid injury or recover if he or she is injured. The U.S. Army is very serious about this effort. It has already spent $50 million creating a Center for Soldier Nanotechnologies at MIT, and it is investing a great deal of money in other research efforts as well.

**Camouflage**

Nanotechnology will directly affect the war fighter at all levels. One good example of what the future may hold is the uniform. *Combat fatigues* or *BDUs* are one of the most overlooked parts of a soldier's kit. They are dyed with camouflage patterns and the fabric is resilient, but there is little else to recommend them. When we discussed weapons of mass destruction, we introduced the idea of making the uniform into a matrix to which sensors could be attached. This is only the beginning.

Protecting a soldier is important, but ideally he or she should also be difficult to see and therefore difficult to attack. This is the idea behind current camouflage uniforms, stealth bombers, and the ability to fight effectively at night—advanced night vision gear is already based on nanotechnology and represents an early application.

The *camouflage* in current uniforms is called *passive camouflage*, meaning that it doesn't change or adapt to different circumstances. *Active camouflage*, however, is just the opposite—it blends in with whatever environment it is exposed to, just as a *chameleon blends* in with rocks, leaves, or soil. A chameleon like approach to making soldiers very difficult to detect may be possible through nanotechnology and *biomimicry*. By working at the same scale as nature, natural processes like those of the chameleon can be emulated and integrated into useful devices. James Bond fans recently saw an exaggerated version of active camouflage in *Die Another Day* in which 007's car could disappear into the ice with the flick of a switch. While active camouflage probably won't be good enough to fool someone up close, active camouflage could certainly have applications for jets and aircraft, making them as hard to see with the eye as they are hard to see with radar.

While it is best to avoid injury through stealth wherever possible, it is necessary to minimize the damage a soldier sustains in an attack should one occur. One approach to doing this involves improving Kevlar the fabric used to make most bulletproof vests. Kevlar is a polymer whose shape is rigid as opposed to the more pliable, thread-like structure of most polymer chains. This rigidity allows Kevlar to disperse energy from local impacts pretty well and to prevent bullets from getting through. Very new smart materials being developed by Ray Baughman's group at the University of Texas in Dallas incoporate nanotube fibers into an open-weave cloth that is more than four times tougher than spider silk and seventeen times tougher than Kevlar.

Researchers at MIT are pursuing another approach. This involves exploring the potential of making a uniform that not only reacts to chemical or biological toxins, but can stiffen and act as armor against ballistic threats such as bullets and fragmentation. Imagine a sheet of paper held at one end. It is floppy and flexible, but if you curve it slightly as you hold it, it becomes stiff and rigid. One approach to allowing a uniform to act as armor involves packing nano particles of iron onto the fabric's fibers. Normally these particles do little other than add a bit of weight, but in the presence of a magnetic field they align and force the fibers to become rigid. The rigid fibers could offer significant protection against ballistic threats and the armor could be activated or deactivated for convenience, depending on whether the soldier is in battle or just patrolling. While iron nano particles may add a bit of weight, they will be significantly lighter than the 15-pound vests that are common today.

Unfortunately, while improved armor will reduce injuries, even nano-protected soldiers can still be hurt in battle. In recent conflicts, some 50% of battlefield fatalities were not immediate but were the result of bleeding or complications occurring several minutes, hours, or even days after a wound was inflicted. Nanotechnology may help to reduce this problem.

If a soldier is hurt in the middle of a firefight, he or she must wait for some kind of assistance from a corpsman or a comrade with a medical kit. What if the nano-enhanced uniform could help with this, too? Sensors could be woven into the fabric that could look for hemoglobin, the oxygen-carrying compound in the blood. Good hemoglobin sensors have already been developed for medical diagnostics such as the detection of blood in stool samples, and these could be incorporated into labs-on-chips and used to begin treatment. Along with the hemoglobin sensors, the tiny labs could also contain supplies of antiseptics, antibiotics, and anesthetics. The fabric itself could contain a smart material liner that would adhere to the wound and act as a temporary bandage.

Groups like the Institute for New Materials in Germany have already demonstrated using *biocidal nanoparticles* (particles that kill bacteria on contact) to keep medical equipment such as hearing aids sterile. These nano particles aren't made of anything exotic—silver is one of the best nano-biocides. It has been suggested that these kinds of biocidal particles could be incorporated into hospital bed sheets to reduce the incidence of cross-infection, and they could be used in uniforms too. Wearing a biocidal uniform could help reduce infection from injuries as well as reduce other kinds of contamination that can occur when uniforms cannot be changed or cleaned regularly. It is not uncommon for a soldier in the battlefield to go for several days working hard, sweating and perhaps bleeding, without having an opportunity to change clothes or visit the laundry.

We began this section with a criticism of a *soldier's kit*. We described it as too cumbersome, and too bulky. We accused it of a lack of integration. Yet now we are suggesting adding a wire mesh, hundreds of sensors, iron nano particles for protection, silver nanoparticles for the reduction of infection, nanoscale medical kits, and who knows what to enable *active camouflage*. It sounds like we've made the situation rather worse than better, but this is not the case.

First of all, these technologies are not as unrelated as they seem. They should all be integrated into a single, coherent system—the next generation uniform. This system will contain a central communications channel in the form of a wire mesh or similar solution and integrated processors to handle information. It will contain functional components or subsystems that react to specific threats. Many of these subsystems have multiple applications. A sensor can be used to detect an external threat, but it may also be used to detect blood and apply treatment. Clothing that can change from flexible to stiff may function as armor or as a cast or tourniquet for medical purposes. One approach to integrating all of these solutions into a uniform involves placing nanoscale channels into the core of the fibers within the fabric. Not only would this solution enable the applications we've discussed, but the electromechanical channel opens the door to using the uniform to actually enhance human performance. It might, for example, augment human muscles and make the wearer effectively stronger, thus providing a unique solution to the weight problem.

Whether they enhance performance or not, weight may not be an issue for such a uniform. To understand why, remember just how small the nanoscale is. Ask anyone who has worn a pair of Nano Care pants if there is any difference in weight between them and ordinary khakis. Ask anyone who has used a Wilson Double-Core tennis ball (based on nanotechnology enhanced composites) if there is any difference in weight between them and ordinary tennis balls. See if you can feel the difference in weight between a microchip with copper technology and one based on aluminum. While it isn't certain, it is likely that one or more of the improvements we have talked about will be possible without affecting the weight or comfort a uniform except in a positive direction.

Ground forces tend to remain in trouble spots long after the Air Force, Navy, and Marines have pulled out. They act as police, peacekeepers and even engineers in the restoration of basic services after fighting is over. They need a better layer of lightweight protection to help them get these jobs done. Nanotechnology provides the only likely solutions to these problems.

**Superhuman**

The tasks of modern soldiers might well be called *superhuman* and thus require superhuman characteristics to accomplish them. We briefly mentioned using channels in the fibers of uniforms to provide performance enhancement for soldiers, but a lot of other approaches are possible as well.

One performance enhancement possibility for increasing strength and reflexes, as well as flying aircraft or driving cars, is a direct interface between the human nervous system and electronics. There have been many promising results in this direction, particularly in the area of prosthetics and bionics. Hearing aids, for example used to work by simply amplifying the sounds around them. *Neuro-electronic hearing aids* work by actually transforming (technically the term is transducing) the sound into a neural impulse and injecting it into the nervous system bypassing the eardrum entirely. Similarly, neuro-electronic eyes attempt to project images directly onto the retina. These devices were created to solve the most untreatable forms of blindness and deafness and are having good results. Optobionics, a company that develops vision solutions, has demonstrated that some trial patients who could see only darkness now see blurry shapes and those who saw blurry shapes can now distinguish between teams at a ball game.

This kind of interface promises treatments for all sorts of conditions from missing limbs to quadriplegia. As it becomes less invasive it could be used to eliminate the old-fashioned steering yoke so pilots can fly planes directly (giving the expression "*fly-by-wire*" a whole new meaning) or to control servo-motors in a battle suit which could increase speed, strength, protection, and more. Working at the scale of nature will make this possible.

## Unmanned Vehicle

Perhaps one of the most controversial ideas for protecting soldiers involves taking them out of combat. This does not mean that wars and conflicts can be entirely avoided, but it may be possible to use more and more mechanization to fight future wars.

Consider the Predator UAV (*Unmanned Aerial Vehicle*). The Predator is just one in a family of unmanned aircraft produced by General Atomics, but in 2001 it made history as the first unmanned aircraft to launch a live antitank missile in combat. While not strictly based on nanotechnology, the Predator shows an interesting trend. Remotely controlled combat aircraft are now a reality as are autonomous missile systems like the Tomahawk cruise missile, which can find a target using GPS and other programming even over complex terrain. Indeed, missiles guided by lasers and heat are now old hat.

Ground vehicles, however are not yet automated. Unlike Predator or Tomahawk, the missions of unmanned ground vehicles do not involve simply going to a location, attempting a kill, and, in the case of Predator, returning to base. A tank or other ground combat vehicle must negotiate changing terrain and deal with rapidly changing threats. Even the short period of latency (the time between when a signal is sent and received) between it and a remote controller could make a world of difference. This is very similar to the set of challenges NASA faces in creating landing vehicles for Mars. To be successful, it will have to be somewhat autonomous, and it will likely rely on several forms of information technology and processing that are not yet fully developed. These include image recognition and, more controversially, artificial intelligence.

It may be a long time before the military or the public is ready for unmanned vehicles or even robots accompanying soldiers into battle, but DARPA (the *Defense Advanced Research Projects Agency*) and others have been looking at the problem for some years.

Several of the components required to make autonomous land vehicles will come from nanotechnology. These include high-performance computers using new architectures that may make artificial intelligence a reality. This artificial intelligence will not necessarily resemble human intelligence, but it will more likely be a specialized system capable of making intelligent decisions of a certain type, a sort of expert system. Thus far even the most advanced silicon computers have been able to show only modest results in this regard since they are designed on an architecture optimized for numerical computation, not cognitive intelligence. Nanotechnological approaches to computing including the use of molecular electronics, bioelectronics (electronic components integrating proteins and other biological entities), and quantum computing may change this.

## Speedy Robots

While autonomous robotic land vehicles may still be in the future, nanotechnology has changes in store for conventional combat vehicles, such as fighters, helicopters, and tanks. Most of these changes will come from the application of nanotechnology-based smart materials.

For example, consider stealth aircraft. While the most advanced models are said to have radar profiles comparable to that of a bumblebee, they make large sacrifices in terms of design. The first stealth fighter, the F-117, has an exotic, panel-shaped design and is far from being the most maneuverable or speedy fighter in the sky. The F/A-22 and other later stealth designs have improved on the F-117, and nanotechnology will help still more. Nanoparticles with massive absorption cross sections for radar and infrared are being developed. This means that they soak tip a huge amount of energy and can therefore prevent detection. In addition, the same active camouflage we discussed for soldiers' uniforms may be applied to aircraft and ground vehicles.

Nanotechnology will also make vehicles faster. Argonide Corporation already produces *Alex*, a rocket-fuel additive based on nano particles of aluminum. *Alex* vastly increases the efficiency of burning hydrocarbon fuels like kerosene and 0 have applications for jet fuels. Such an additive could also increase the efficiency of powder burn in a rifle cartridge.

Another barrier to speed, fuel efficiency, and deployment is the weight of combat vehicles. A tank may weigh 60 tons and infantry vehicles weigh almost as much. Lighter materials such as the ever-

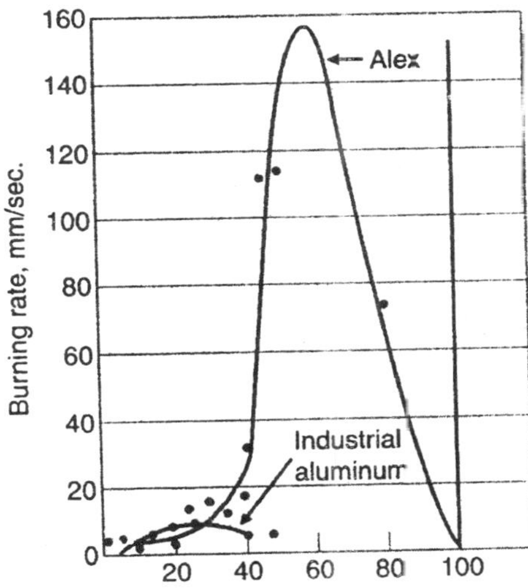


*Fig. 18.1. This chart shows the relative performance of Alex, a nanopowdered fuel additive, and its microparticle competitors.*

useful carbon nano tube may help solve this problem. Next-generation commercial aircraft already utilize car bon fiber and other super strong, lightweight carbon composites in their construction, but none of these materials can come close to the prowess of nanotubes. With strength on the order of 60 times that of steel at a fraction of the weight, nanotubes will be used to make aircraft that are faster and that require shorter runways and less fuel. Nanotubes or other such nanomaterials can even be used to armor aircraft such as attack helicopters which are usually lightly armored for weight reasons. Companies like CNI and NanoLab are now offering nanotubes in appreciable quantities, but they still can cost about a hundred dollars a gram. Still, this price is far lower than it was just two years ago, and two years before that the idea of bulk nano tubes was still a pipe dream.

While *carbon nanotubes* may not yet be the armor of choice, nano-enhanced armor is already in use protecting troops. The SAG (Save a Gunner) turret, for light combat vehicles such as HMMWVs is a case in point. Produced by US Global Nanospace, it uses nanofiber materials to create a gun turret that is less than 200 pounds, has minimal impact on mobility, and yet can withstand six successive 7.62-caliber bullet strikes in the same spot. US Global Nanospace uses similar technology to make blast-proof doors for aircraft and materials for dealing with explosives.

NASA engineers have started thinking about what its next-generation orbiter might look like, and it also incorporates many of these ideas. In addition, NASA scientists propose using nanotechnology-based coatings to make the skin of a new spacecraft almost friction free, reducing heat from reentry. They are also considering the idea of self-healing materials—these are materials that can start to repair themselves if they are damaged in the same way that human skin does. Self-healing materials often operate on a similar premise to skin. Some of the first self-healing plastics have been developed at the University of Illinois. They work by incorporating microspheres of liquid monomers (the building blocks of plastic polymers) and a catalyst into the plastic matrix. These micro- spheres correspond to platelets in human blood. When a crack forms in the plastic the micro spheres break to expose the monomers and the catalyst, which then move in to seal the crack just as platelets move in to seal wounds in skin. Self-healing materials on aircraft could possibly have prevented the crash of American Airlines flight 587 in New York, the crash of Air France Concorde flight 4590 in Paris, and even the explosion of the space shuttle Columbia. Aside from their applications for civil aviation, they could also make combat aircraft much more survivable.

## Alternative Energy Sources

Alternative energy sources will be necessary to take advantage of many of these new nanotechnologies. Current equipment is already power-hungry, and many of the technologies we have discussed will exacerbate the problem. Soldiers currently carry more than 20 pounds of batteries and portable power supplies into battle, but consider the issues of other kinds of combat equipment. Ground vehicles and aircraft are powered using hydrocarbon fuels (usually diesel or kerosene, although some multi-fuel engines can also run on gasoline or even alcohol). These fuel supplies are very volatile and can result in "secondary explosions," blasts that occur after an attack usually as a result of hazardous cargo such as fuel or ammunition. Fuel depots represent huge targets for the enemy and transporting fuel is a logistical nightmare.

Naval vessels have dealt with this problem in many cases by using nuclear fission reactors. These have the advantage that they seldom need to be refueled (only every few years) and thus give the vessels effectively unlimited operational range, but they also create other hazards. If a nuclear vessel is disabled, the fissionable material may be recovered and used for making a bomb. If it is destroyed or damaged the radioactive materials it contains could cause an environmental catastrophe.

Promising nanotechnology-based energy solutions such as fuel cells, efficient solar power, and even nuclear fusion (which has a nano component) could reduce or eliminate these concerns.

## Nanoarchitecture

The applications we have discussed in this chapter represent only the beginning of the possible applications of nanotechnology for defense. Nanotechnology will greatly increase the survivability of soldiers and equipment, especially for peacekeepers in the period after the hottest fighting is over and when a full combat rig is impractical. It will provide stealthier, tougher and faster combat vehicles and increase battle space information and awareness. It will improve communication within Units and allow a soldier's kit to become an efficient, integrated system. It will combat the threats of chemical an biological weapons. It will be a crucial tool for weapon inspectors and those looking for weapons of mass destruction. It will provide lighter; more efficient portable power. It already helps our troops to see at night and provides some of the few chemical and biological defenses that we have so far developed. For these reasons, nanotechnology is key to the military's new mission.

When he developed one of the first machine guns during the Civil War, Richard Gatling sincerely hoped that the destruction it would wreak would be so great that no one would dare to use it and wars would become impossible. Clearly this did not happen, and the advice that guided one of his successors, Hiram Maxim, seemed much more apt: "If you wanted to make a lot of money, invent something that will enable these Europeans to cut each other's throats with greater facility." Perhaps Gatling's vision was most nearly attained by the nuclear weapon, but even the MAD doctrine depends on sane and stable governments having exclusive control of weapons of mass destruction There are no credible claims that nanotechnology will take the final step in making war obsolete, hut one thing that makes nanotechnology particularly exceptional among technologies with military applications is that it is primarily useful for defense. It can enhance soldier safety and capabilities, limit the destruction necessary to accomplish objectives, provide information and intelligence, coordinate troops, enhance reaction times, and even help heal the wounded. It may also make a lot of money. While it certainly could be used for weapons, nuclear weapons are still the ultimate killing machine. Since the American military already has destructive weapons of almost limitless capability, what it needs now is better defense and better capability for using its offensive weapons effectively. That is where nanotechnology will start.

# 19

# FINDING THE FACTS

Society needs better ways to understand technology—this has long been obvious. The challenges ahead simply make our need more urgent. The promise of technology lures us onward, and the pressure of competition makes stopping virtually impossible. As the technology race quickens, new developments sweep toward us faster, and a fatal mistake grows more likely. We need to strike a better balance between our foresight and our rate of advance. We cannot do much to slow the growth of technology, but we can speed the growth of foresight. And with better foresight, we will have a better chance to steer the technology race in safe directions.

Various approaches to guiding technology have been suggested. "The people must control technology" is a plausible slogan, but it has two possible meanings. If it means that we must make technology serve human needs, then it makes good sense. But if it means that the people as a whole must make technical decisions, then it makes very little sense. The electorate cannot judge the intricate links between technology, economy, environment, and life; people lack the needed knowledge. The people themselves agree: according to a U.S. National Science Foundation survey, 85 percent of U.S. adults believe that most citizens lack the knowledge needed to choose which technologies to develop. The public generally leaves technical judgments to technical experts.

Unfortunately, leaving judgment to experts causes problems. In *Advice* and *Dissent*, Primack and von Hippel point out that "to the extent that the Administration can succeed in keeping unfavourable information quiet and the public confused, the public welfare can be sacrificed with impunity to bureaucratic convenience and private gain." Regulators suffer more criticism when a new drug causes a single death than they do when the *absence* of a new drug causes a thousand deaths. They misregulate accordingly. Military bureaucrats have a vested interest in spending money, hiding mistakes, and continuing their projects. They mismanage accordingly. This sort of problem is so basic and natural that more examples are hardly needed. Everywhere, secrecy and fog make bureaucrats more comfortable; everywhere, personal convenience warps factual statements on matters of public concern. As technologies grow more complex and important, this pattern grows more dangerous.

Some authors consider rule by secretive technocrats to be virtually inevitable. In *Creating Alternative Futures*, Hazel Henderson argues that complex technologies "become inherently totalitarian" (her italics) because neither voters nor legislators can understand them. In The *Human Future Revisited*, Harrison Brown likewise argues that the temptation to bypass democratic processes in solving complex crises brings the danger "that if industrial civilization survives it will become increasingly totalitarian in nature." If this were so, it would likely mean our doom: we cannot stop the technology race, and a world of totalitarian states based on advanced technology needing neither workers nor soldiers might well discard most of the population.

Fortunately, democracy and liberty have met comparable challenges before. States grew too complex for direct democracy, but representative government evolved. State power threatened to crush liberty, but the rule of law evolved. Technology has grown complex, but this gives us no reason to ignore the people, discard the law, and hail a dictator. We need ways to handle technical complexity in a democratic framework, using experts as instruments to clarify our vision without giving them control of our lives. But technical experts today are mired in a system of partisan feuding.

### ESTABLISHMENT

Government and industry—and their critics—commonly appoint expert committees that meet in secret, if they meet at all. These committees claim credibility based on *who they are*, not on *how they work*. Opposed groups recruit opposed Nobel laureates.

To gain influence in our mass democracy, groups try to outshout one another. When their views have corporate appeal, they take them to the public through advertising campaigns. When their views have pork-barrel appeal, they take them to legislatures through lobbying.

When their views have dramatic appeal, they take them to the public through media campaigns. Groups promote their pet experts, the battle goes public, and quiet scientists and engineers are drowned in the clamor.

As the public conflict grows, people come to doubt expert pronouncements. They judge statements the obvious way, by their source. ("Of course she claims oil spills are harmless—she works for Exxon." "Of course he says Exxon lies—he works for Nader.")

When established experts lose credibility, demagogues can join the battle on an equal footing. Reporters—eager for controversy, striving for fairness, and seldom guided by technical backgrounds-carry all sides straight to the public. Cautious statements by scrupulous scientists make little impression; other scientists see no choice but to adopt the demagogues' style. Debates become sharp and angry, divisions grow, and the smoke of battle obscures the facts. Paralysis or folly often follows.

Our greatest problem is how we handle problems. Debates rage over the safety of nuclear power, coal power, and chemical wastes. Well-meaning groups backed by impressive experts clash again and again over dull, technical facts—dull, that is, save for their importance: What are the effects of low-level radiation, and how likely is a reactor meltdown? What are the causes and effects of acid rain? How well could space-based defenses block missile attacks? Do five cases of leukemia within three miles of a particular waste dump show a deadly hazard, or merely the workings of chance?

Greater issues lie ahead: How safe is this replicator? Will this active shield system be safe and secure? Will this biostasis procedure be reversible? Can we trust this AI system?

Disputes about technical facts feed broader disputes about policy. People may have differing values (which would you rather have, encephalitis or pesticide poisoning?) but their views of relevant facts often differ still more. (How often do these mosquitoes carry encephalitis? How toxic is this pesticide?) When different views of boring facts lead to disagreements about important policies, people may wonder, "How can they oppose us on this vital issue unless they have bad motives?" Disputes over facts can thus turn potential allies against one another. This hampers our efforts to understand and solve our problems.

People have disputed facts for millennia; only the prominence of technical disputes is new. Societies have evolved methods for judging facts about people. These methods suggest how we might judge facts about technology.

## Trial and Hit

Throughout history, groups have evolved ways to resolve disputes; the alternative has been feuds, open-ended and often deadly. Medieval Europeans used several procedures, all better than endless feuding: They used trial by battle: opponents fought, and the law vindicated the victor.

They used compurgation: neighbours swore to the honesty of the accused; if enough swore, the charges were dropped. They used trial by ordeal: in one, the accused was bound and thrown in a river; those who sank were innocent, those who floated, guilty.

They used judgment by secretive committees: the king's councilors would meet to judge and pass sentence as seemed fit. In England, they met in a room called the Star Chamber.

These methods supposedly determined *who did what*—the facts about human events—but all had serious shortcomings. Today we use similar methods to determine *what causes what*—the facts about science and technology:

We use trial by combat in the press: opponents fling sharp words until one side's case suffers political death. Unfortunately, this often resembles an endless feud.

We use compurgation: experts swear to certain facts; if enough swear the same, the facts are declared true. We use judgment by secretive committees: selected experts meet to judge facts and recommend such actions as seem fit. In the United States, they often meet in committees of the National Academy of Sciences.

Trial by ordeal has passed from fashion, but combat in the press may well seem like torture to the quiet scientist with self-respect.

The English abolished Star Chamber proceedings in 1641, and they counted this a great achievement. Trial by combat, compurgation, and ordeal have likewise become history. We now value due process, at least when judging people.

Court procedures illustrate the principles of due process: Allegations must be specific. Both sides must have a chance to speak and confront each other, to rebut and cross-examine. The process must be public, to prevent hidden rot. Debate must proceed before a jury that both sides accept as impartial. Finally, a judge must referee the process and enforce the rules. To see the value of due process, imagine its opposite: a process trampling all these principles would give one side a say and the other no chance to cross-examine or respond. It would meet in secret, allow vague smears, and lack a judge to enforce whatever rules might remain. Jurors would arrive with their decisions made. In short, it would resemble a lynch mob meeting in a locked barn—or a rigged committee drafting a report.

Experience shows the value of due process in judging facts about people; might it also be of value in judging facts about science and technology? Due process is a basic idea, not restricted to courts of law. Some AI researchers, for example, are building due-process principles into their computer programs. It seems that due process should be of use in judging technical facts. In fact, the scientific literature—the chief forum of science—already embodies a form of due process: In good journals, scientific statements must be specific. In theory, given enough time and persistence, all sides may state their views in a dispute, since journals stand open to controversy. Though opponents may not meet face to face, they confront each other at a distance; they question and respond in slow motion, through letters and articles. Referees, like juries, evaluate evidence and reasoning. Editors, like judges, enforce rules of procedure. Publication keeps the debate open to public scrutiny.

In both journals and courts, conflicting ideas are pitted against one another under rules evolved to ensure a fair, orderly battle. These rules sometimes fail because they are broken or inadequate, but they are the best we have developed. Imperfect due process has proved better than no due process at all.

Why do scientists value refereed journals? Not because they trust all refereed journals, or trust everything printed in any one of them. Even the best due-process system won't grind out a stream of pure truth. Rather, they value refereed journals because they tend to reflect sound critical discussion. Indeed, they must: because journals compete with one another for papers, prestige, and readership, the best journals must be good indeed. Journals grind slowly, yet after enough rounds of publication and criticism they often grind out consensus. Experience proves the value of both courts and journals. Their underlying similarity suggests that their value stems from a common source—due process. Due process can fail, but it is still the best approach known for finding the facts.

Today, courts and journals are not enough. Vital technical disputes go on and on because we have no rapid, orderly way to bring out the facts (and to delineate our ignorance). Courts are not suited to deal with technical questions. Journals are better, but they still have shortcomings. They took shape in a time of lower technology and slower advance, evolving to fit the limits of printing, the speed of mail, and the needs of academic science. But today, in a time when we desperately need better and swifter technical judgment, we find ourselves in a world that has telephones, jets, copiers, and express and electronic mail. Can we use modern technologies to speed technical debate?

Of course: scientists already use several approaches. Jets bring scientists from around the globe to conferences where papers are presented and discussed. But conferences handle controversy poorly; public decorum and tight schedules limit the vigor and depth of debate. Scientists also join informal research networks linked by telephone, mail, computers, and copying machines; these accelerate exchange and discussion. But they are essentially private institutions. They, too, fail to provide a credible, public procedure for thrashing out differences.

Conferences, journals, and informal networks share some similar limitations. They typically focus on technical questions of scientific importance, rather than on technical questions of public-policy importance. Moreover, they typically focus on *scientific* questions. Journals tend to slight technological questions that lack intrinsic scientific interest; they often treat them as news items not worthy of checking by referees. Further, our present institutions lack any balanced way to present knowledge when it is still tangled in controversy. Though scientific review articles often present and weigh several sides, they do so from a single author's point of view.

All these shortcomings share a common source: scientific institutions evolved to advance science, not to sift facts for policy makers. These institutions serve their purpose well enough, but they serve other purposes poorly. Though this is no real failing, it does leave a real need.

## Better Procedure

We need better procedures for debating technical facts—procedures that are open, credible, and focused on finding the facts we need to formulate sound policies. We can begin by copying aspects of other due-process procedures; we then can modify and refine them in light of experience. Using modern communications and transportation, we can develop a focused, streamlined, journal-like process to speed public debate on crucial facts; this seems half the job. The other half requires distilling the results of the debate into a balanced picture of our state of knowledge (and by the same token, of our state of ignorance). Here, procedures somewhat like those of courts seem useful.

Since the procedure (a *fact forum*) is intended to summarize facts, each side will begin by stating what it sees as the key facts and listing them in order of importance. Discussion will begin with the statements that head each side's list. Through rounds of argument, cross examination, and negotiation the referee will seek agreed-upon statements. Where disagreements remain, a *technical panel* will then write opinions, outlining what seems to be known and what still seems uncertain. The output of the fact forum will include background arguments, statements of agreement, and the panel's opinions. It might resemble a set of journal articles capped by a concise review article - one limited to factual statements, free of recommendations for policy. This procedure must differ from that of a court in various ways. For example, the technical panel—the forum's "jury"—must be technically competent. Bias might lead a panel to misjudge facts, but technical incompetence would do equal harm. For this reason, the "jury" of a fact forum must be selected in a way that might be dangerous if allowed in courts of law. Since courts wield the power of the police, we use juries selected from the people as a whole to guard our liberty.

This forces the government to seek approval from a group of citizens before it punishes someone, thus tying the government's actions to community standards. A fact forum, however, will neither punish people nor make public policy. The public will be free to watch the process and decide whether to believe its results. This will give people control enough.

Still, to make a fact forum fair and effective, we will need a good panel-selection procedure. Technical panels will correspond roughly to the expert committees appointed by governments or to the referees appointed by journals. To ensure fairness, a panel must be selected not by a committee, a politician, or a bureaucrat, but by a process that involves the consent of both sides in the dispute. In court proceedings, advocates can challenge and reject any jurors who seem biased; we can use a similar process in selecting the panel for a fact forum. Experts who are directly involved in a dispute can't serve on the panel - they would either bias the panel or split it. The group sponsoring a fact forum must seek panelists who are knowledgeable in related fields. This seems practical because the methods of technical judgment (often based on experiments and calculations) are quite general. Panelists familiar with the fundamentals of a field will be able to judge the detailed arguments made by each side's specialists.

Other parts of the fact forum will also resemble those of courts and journals. A committee like a journal's editorial group will nominate a referee and panelists for a dispute. Advocates for each side, like authors or attorneys, will assemble and present the strongest case they can.

Despite these similarities, a fact forum will differ from a court: It will focus on technical questions. It will suggest no actions. It will lack government power. It will follow technical rules of evidence and argument. It will differ in endless details of tone and procedure. The analogy with a court is just that - an analogy, a source for ideas.

A fact forum will also differ from a journal: It will move as fast as mail, meetings, and electronic messages permit, rather than delaying exchanges by many months, as in typical journal publication. It will be convened around an issue, rather than being established to cover a scientific field. It will summarize knowledge to aid decisions, rather than serving as a primary source of data for the scientific community. Although a series of fact forums won't replace a journal, they will help us find and publicize facts that could save our lives. Dr. Arthur Kantrowitz (a member of the National Academy of Sciences who is accomplished in fields ranging from medical technology to high-power lasers) originated the concept I have just outlined. He at first called it a "board of technical inquiry." Journalists promptly dubbed it a "science court." I have called it a "fact forum"; I will reserve the term "science court" for a fact forum used (or proposed) as a government institution. Proposals for due process in technical disputes are still in flux; different discussions use different terms.

Dr. Kantrowitz's concern with due process arose out of the U.S. decision to build giant rockets to reach the Moon in one great leap; he, backed by the findings of an expert committee, had recommended that NASA use several smaller rockets to carry components into a low orbit, then plug them together to build a vehicle to reach the Moon. This approach promised to save billions of dollars and develop useful space-construction capabilities as well. No one answered his arguments, yet he failed to win his case. Minds were set, politicians were committed, the report was locked in a White House safe, and the debate was closed. The technical facts were quietly suppressed in the interests of those who wanted to build a new generation of giant rockets.

This showed a grave flaw in our institutions—one that persists, wasting our money and increasing the risk of a disastrous error. Dr. Kantrowitz soon reached the now obvious conclusion: we need due process institutions for airing technical controversies.

Dr. Kantrowitz pursued this goal (in its science-court form) through discussions, writings, studies, and conferences. He won endorsements for the science court idea from Ford, Carter, and Reagan—as candidates. As Presidents, they did nothing, though a presidential advisory task force during the Ford administration did detail a proposed procedure.

Still, progress has been made. Although I have used the future tense in describing the fact forum, experiments have begun. But before describing a path to due process, it makes sense to consider some of the objections.

## Facts and Critics

Critics of this idea (at least in its science court version) have often disagreed with one another. Some have objected that factual disputes are unimportant, or that they can be smoothed over behind closed doors; others have objected that factual disputes are too deep and important for due process to help. Some have warned that science courts would be dangerous; others have warned that they would be impotent. These criticisms all have some validity: due process will be no cure-all. Sometimes it won't be needed, and sometimes it will be abused. Still, one might equally well reject penicillin on the grounds that it is sometimes ineffective, unnecessary, or harmful.

These critics propose no alternatives, and they seldom argue that we have due process today, or that due process is worthless. We must deal with complex, technical issues on which millions of lives depend; dare we leave these issues to secretive committees, sluggish journals, media battles, and the technical judgment of politicians? If we distrust experts, should we accept the judgment of secretive committees appointed in secret, or demand a more open process? Finally, can we with our present system cope with a global technology race in nanotechnology and artificial intelligence?

Open, due-process institutions seem vital. By letting all sides participate, they will harness the energy of conflict to a search for the facts. By limiting experts to describing the facts, they will help us cope with technology without surrendering our decisions to technocrats. Individuals, companies, and elected officials will keep full control of policy; technical experts will still be able to recommend policies through other channels.

How can we distinguish facts from values? Karl Popper's standard seems useful: a statement is factual (whether true or false) if an experiment or observation could in principle disprove it. To some people, the idea of *examining facts* without considering values suggests the idea of *making policy* without considering values. This would be absurd: by their very nature, policy decisions will always involve both facts and values. Cause and effect are matters of fact, telling us what is possible. But policy also involves our values, our motives for action. Without accurate facts, we won't get the results we seek, but without values - without desires and preferences—we wouldn't seek anything in the first place. A process that uncovers facts can help people choose policies that will serve their values.

Critics have worried that a science court will (in effect) declare the Earth to be flat, and then ignore an Aristarchus of Samos or a Magellan when he finds evidence to the contrary. Errors, bias, and imperfect knowledge will surely cause some memorable mistakes. But members of a technical panel need not claim a bogus certainty. They can instead describe our knowledge and outline our ignorance, sometimes stating that we simply don't know, or that present evidence gives only a rough idea of the facts. This way, they will protect their reputations for good judgment. When new evidence arrives, a question can he reopened; ideas need no protection from double jeopardy.

If fact forums become popular and respected, they will gain influence. Their success will then make them harder to abuse: many competing groups will sponsor them, and a group that abuses the procedure will tend to gain a bad reputation and be ignored. No single sponsoring group will be able to obscure the facts about an important issue, if fact forums gain reliability through redundancy.

No institution will be able to eliminate corruption and error, but fact forums will be guided, however imperfectly, by an improved standard for the conduct of public debate; they will have to fall a long way before becoming worse than the system we have today. The basic case for fact forums is that (1) due process is the right approach to try, and (2) we will do better if we try than if we don't.

## Public Benefits

Anthropologist Margaret Mead was invited to a colloquium on the science court to speak against the idea. But when the time came, she spoke in its favour, remarking that "We need a new institution. There isn't any doubt about that. The institutions we have are totally unsatisfactory. In many cases, they are not only unsatisfactory, they involve a prostitution of science and a prostitution of the decision making process." People with no vested interest in the existing institutions often agree with her evaluation.

If finding the facts about technology really is crucial to our survival, and if due process really is the key to finding the facts, then what can we do about it? We needn't begin with perfect procedures; we can begin with informal attempts to improve on the procedures we have. We can then evolve better procedures by varying our methods and selecting those that work best. Due process is a matter of degree.

Existing institutions could move toward due process by modifying some of their rules and traditions. For example, government agencies could regularly consult opposing sides before appointing the members of an expert committee. They could guarantee each side the right to present evidence, examine evidence, and cross-examine experts. They could open their proceedings to observers. Each of these steps would strengthen due process, changing Star Chamber proceedings into institutions more worthy of respect.

The public benefits of due process won't necessarily make it popular among the groups being asked to change, however. We haven't heard the thunder of interest groups rushing to test their claims, nor the cries of joy from committees as they throw open their doors and submit to the discipline of due process. Nor have we heard reports of politicians renouncing the use of spurious facts to hide the political basis of their decisions.

Yet three U.S. presidential candidates did endorse science courts. The Committee of Scientific Society Presidents, which includes twenty-eight of the leading scientific societies in the U.S., also endorsed the idea. The U.S. Department of Energy used a "science court-like procedure" to evaluate competing fusion-power proposals, and declared it efficient and useful. Dr. John C. Bailar of the National Cancer Institute, after failing to make medical organizations recognize the dangers of X rays and reduce their use in mass screening, proposed holding a science court on the subject. His opponents then backed down and changed their policies—apparently, the mere threat of due process is already saving lives. Nevertheless, the old ways continue almost unchallenged. Why is this? In part, because knowledge is power, and hence jealously guarded. In part, because powerful groups can readily imagine how due process would inconvenience them. In part, because an effort to improve problemsolving methods lacks the drama of a campaign to fight problems directly; a thousand activists bail out the ship of state for every one who tries to plug the holes in its hull.

Governments may yet act to establish science courts, and any steps they may take toward due process merit support. Yet it is reasonable to fear government sponsorship of science courts: centralized power tends to beget lumbering monsters. A central "Science Court Agency" might work well, do little visible harm, and yet impose a great hidden cost: its very existence (and the lack of competition) might block evolutionary improvements. Other paths lie open. Fact forums will he able to exert influence without help from special legal powers. To have a powerful effect, their results only need to be more credible than the assertions of any particular person, committee, corporation, or interest group. A well-run fact forum will carry credibility in its very structure. Such forums could be sponsored by universities. In fact, Dr. Kantrowitz recently conducted an experimental procedure at the University of California at Berkeley. It centered around public disputes between geneticist Beverly Paigen and biochemist William Havender regarding birth defects and genetic hazards at the Love Canal chemical dump site. They served as advocates; graduate students served as a technical panel. The procedure involved meetings spread over several weeks, chiefly spent in discussing areas of agreement and disagreement; it wound up with several sessions of public cross-examination before the panel. Both the advocates and the panel agreed to eleven statements of fact, and they clarified their remaining disagreements and uncertainties.

Arthur Kantrowitz and Roger Masters note that "in contrast to the difficulties experienced in the many attempts to implement a science court under government auspices, encouraging results were obtained in the first serious attempt... in a university setting." They remark that the traditions and resources of universities make them natural settings for such efforts. They plan more such experiments.

This shows a decentralized way to develop due-process institutions, one that will let us outflank existing bureaucracies and entrenched interests. In doing this, we can build on established principles and test variations in the best evolutionary tradition.

Leaders concerned with a thousand different issues—even leaders on opposing sides of an issue—can join in support of due process. In Getting to Yes, Roger Fisher and William Ury of the Harvard Negotiation Project point out that opponents tend to favour impartial arbitration when each side believes itself right. Both sides expect to win, hence both agree to participate. Fact forums should attract honest advocates and repel charlatans. As due process becomes standard, advocates with sound cases will gain an advantage even if their opponents refuse to cooperate—"If they won't defend their case in public, why listen to them?" Further, many disputes will be resolved (or avoided) without the trouble of actual proceedings: the prospect of a refereed public debate will encourage advocates to check their facts before they take stands. Establishing fact forums could easily do more for a sound cause than would the recruitment of a million supporters.

Yet today well-meaning leaders may feel forced to exaggerate their cases simply to be heard above the roar of their opponents' press releases. Should they regretfully oppose the discipline of fact forums? Surely not; due process can heal the social illness that forced them away from the facts in the first place. By making their points in a fact forum, they can recover the selfrespect that comes with intellectual honesty. Unearthing the truth may undermine a cherished position, but this cannot harm the interests of a true publicinterest group. If one must move a bit to make room for the truth, so what? Great leaders have shifted positions for worse reasons, and due process will make opponents shift too.

Gregory Bateson once stated that "no organism can afford to be conscious of matters with which it could deal at unconscious levels." In the organism of a democracy, the conscious level roughly corresponds to debate in the mass media. The unconscious levels consist of whatever processes ordinarily work well enough without a public hue and cry. In the media, as in human consciousness, one concern tends to drive out another. This is what makes conscious attention so scarce and precious. Our society needs to identify the facts of its situation more swiftly and reliably, with fewer distracting feuds in the media. This will free public debate for its proper task—judging procedures for finding facts, deciding what we want, and helping us choose a path toward a world worth living in.

As change quickens, our need grows. To judge the risk of a replicator or the feasibility of an active shield, we will need better ways to debate the facts. Purely social inventions like the fact forum will help, but new social technologies based on computers also hold promise. They, too, will extend due process.

# INDEX